# PULSE CROPS

## Socialization and Market Response

*The Authors*

**Prof. Sankar Kr Acharya,** former Head, Dept. of Agril Extension and Director, Extension Education, Bidhan Chandra Krishi Viswavidyalaya, Mohanpur, WB, born on 6th October, 1960, started his career as Assistant Professor at BCKV in 1988 and has been in teaching, research and extension over 30 years. Research Publication: 217 in National and International Journals. Book Publication: 99 books authored/ co-authored Membership of Distinction: Chancellor Nominee on selection committee for professors' positions; Member of Agriculture Commission; Marketing and Extension Sub Committee, Govt. of West Bengal; Expert member WWF projects in BTR (Buxa Tiger Reserve Project); Expert member; DFID project on Primary Education (DPEP); REIA, Visited Italy, France, Germany, China, Sri Lanka and Bangladesh. Delivered 18 Key note addresses in International Congress, Chaired 19 Sessions in International conferences, Editors of a score of national and international journals. He has so far been awarded with 13 best paper awards in National and International Conferences Organizing Secretary of two International Conferences, He has also been honoured to be selected as the Convener of Panel (PE-32) entitled The hunger , poverty and silence of IUAES, University of Manchester, UK, 2013. He is the fellow, ISEE, IARI, New Delhi.

**Dr. Dibyendu Pal,** Born in 1992, persuaded B.Sc. (Ag.) Hons. from Visva Bharati Viswavidyalaya, and completed his M.Sc. (Ag.) in Agricultural Extension, from Bidhan Chandra Krishi Viswavidyalaya, in Mohan pur, W.B. completed his Ph.D. programme under the guidance of Prof. A. Biswas. As a scholar, he is hard working, innovative and has already developed a mettle and analytical competency on social ecology of rural livelihood as well as participatory agro-ecosystem analysis. He has already participated and presented research papers in couple of National and International conferences.

**Dr. Amitava Biswas,** born in 1964, Professor, Department of Agricultural Extension, BCKV, is an eminent personality in the domain of Extension Science. He has been in teaching, research and extension for the last twenty years in the Department of Agricultural Extension, Bidhan Chandra Krishi Viswavidyalaya, Mohanpur, Nadia, West Bengal. He has built up an incredible acumen in Participatory Extension Methods, Indigenous Technical Knowledge, Rural Entrepreneurship and Developmental studies. So far he has authored 40 research papers in National and International Journals, 19 books and good number of popular articles. He is a good trainer in multi discipline area and regular speaker in TV and Radio. He rendered administrative services as Deputy Director of Extension Education and Editorial Works as Editor of University farm Journal 'Krishi Bidhan'. For the last couple of years he has been with popular Science movement. He has also been involved in some ACRP, NAIP, RKVY and ICARDA Projects.

**Dr. Lakshmi Narsimhaiah,** Assistant Professor (CT), Agricultural Statistics, Department of Basic Science and Humanities, UAHS (Shivamogga), Karnataka is an eminent personality in the domain of Statistics. He received university Gold Medal with first rank for his M.Sc. degree programme in Department of Agricultural Statistics, UAS (B), Bangalore during 2011-13. He also received INSPIRE Fellowship of Department of Science and Technology for pursuing Ph.D. programme at BCKV, Mohanpur, West Bengal during 2014-2019. He served in corporate sector-Tata Consultancy Services in Retail Market Research, Bangalore. He is proficiency at using statistical packages like MS-Excel, SPSS, XL-Stat, Mini Tab, SAS, Easy Fit, R software, Eviews, etc. He is specialized in the field of statistical modelling and forecasting, time series analysis, socio-economic analysis and econometrics. So far, he has authored/ Co-authored twenty plus research articles in national and international Journals.

# PULSE CROPS

## Socialization and Market Response

Sankar Kr Acharya

Dibyendu Pal

Amitava Biswas

Lakshmi Narasimhaiah

2022

**Scholars World**

*A Division of*

**Astral International Pvt. Ltd.**

New Delhi – 110 002

ISBN: 9789354615344



*Published by* : **Scholars World**
*A Division of*
**Astral International Pvt. Ltd.**
– ISO 9001:2015 Certified Company –
4736/23, Ansari Road, Darya Ganj,
New Delhi-110 002
Ph. 011-4354 9197, 2327 8134
E-mail: info@astralint.com
Website: www.astralint.com

**Prof. Baldeo Singh,**

*Ph. D., ARS Former Joint Director Extension, Indian Agricultural Research Institute (IARI), New Delhi, India.*

# FOREWORD

The global agriculture has been passing through a classic contradiction between ecology and economy and, India is no exception to it. The reconciliation between ecology and economy can ensure both ecological and sociological resilience. The degradation of soil organic carbon, loss of top soils, depletion of ground water and gene erosion from our gifted gene pool are being mistakenly underestimated against the flamboyance of yield revolution for major crops. Can we go for an agriculture which disrupts soil, depletes water, pollutes environment and throws the lives of farmers into a vicious cycle of poverty-hunger and silence. Perhaps, Pulse is the only crop on this planet that can cater to all the three primers of evolving agriculture, economy-ecology-technology, on a single go. It adds to nitrogen to soil, protein to human body and income to wallet of farmers, both small and marginal, and people at large. To earn self reliance in pulse production, an array of initiatives has been triggered up to ensure both food and income security for the millions of rural people in India. The book offers the deep insights for delving into and examines the technology socialization process of pulse crops and their market responses in terms of marketable surplus, market disposal, net return, ecological response by the growers after meeting their household needs. An account of 150 pulse growing farmers have been interrogated and interacted with to estimate and interpret the cause, direction and level of socialization of pulse crop at the grass root reality, and thus, hard evidences are extracted therewith from a new alluvial agro-ecological setting of India. These have been extrapolated to a reference book for scholars and academia in agriculture and ecology across the planet.

*Prof. Baldeo Singh*

# PREFACE

The most alarming part of Indian agriculture is that its ecological bases are losing resilience very fast. The plunging of organic carbon status, increasing trend of soil compaction and pleatuing of yield of major food crops are of prime concerns. So addition of pulse crops to the existing cropping sequence has got both ecological and economic implication. We can save million dollars spent on importing pulses by creating of self reliance in pulse production.

The whole of the process needs re-engineering and refocusing our crop plan and market readiness in such a way as to enhance income of farmers, nutrition of the families and resilience of agro-ecosystem. Albeit a reinforced endeavour to scale up production and productivity of both pulse and oilseeds across the nation, its socialization of critical technologies and marketability are not beyond question.

If restoration of ecological properties across the land sizes and categories has been the prime interest, socialization of pulse crops should be scaled up with a community level movement. We hope, the book will offer a great reading worthiness for scholars and faculties across the disciplines.

*S K Acharya*

# CONTENTS

CHAPTER 1

# INTRODUCTION

Pulse provides the green source of protein to millions of Indian and beyond. In India pulse crop have been described as a "poor man's meat and rich man's vegetable". It's a rare type of vegetative protein which retain lysine one of the most important amino acid. As against animal protein, it's the cheaper source of vegetative protein as well. As a crop, it needs less water and nutrient, less cost of investment as well. Indian agriculture cannot fulfil the total pulse requirement, hence, a huge expenditure incurred over pulse import and export.

Pulse is an ecofriendly crop. Pulses are also excellent feed and fodder for livestock. Endowed with the unique ability of biological nitrogen fixation, carbon sequestration, soil amelioration, low water requirement and capacity to withstand harsh climate, pulse crop have remained an integral component of sustainable crop production system since time immemorial, especially in the dry rain fed areas. It adds nitrogen to the soil, retain soil moisture and leave added nutrient for the next crop.

Pulses are important component of diet in Indian, mainly for being one of the leading sources of protein for predominantly vegetarian society. The emphasis of the Government has, accordingly, been on increasing production of pulses through area expansion and adoption of improved technology so as to ensure the availability of pulses to the masses as per quantity recommended by ICMR (Indian Council of Medical Research). Equally important is the marketing of the pulses particularly in a situation created by the introduction of reforms in the sector and existing demand-supply gap in pulses. Shortage of supply of pulses has resulted in increase in prices, thereby pushing pulses out of the reach of poor household leading to a negative effect on their nutritional status (Reddy, 2004). The Indian food economy, since 1970, has also undergone major policy reforms, including trade liberalization leading to opening up of domestic pulse market to international trade. The market is very lucrative and of major significance to the world pulse economy (Agbola & Damoense, 2003) as has been evident by the increased imports of pulses in India in recent past. The marketing component is important to ensure remunerative prices to the farmers which will eventually work as an incentive for them to bring more area under pulses. The Government has provided market price support to the framers through minimum support price of leading pulses. However, the marketing and price policy of the government should be the focus of various government initiatives mainly the extension programmes, so that the awareness may be created among farming community and farmer's may 14 respond more effectively to the various production incentives given by the Government (Savadatti and Narappanavar, 1998). Marketing can also help in inducing an element of incentive to farmer through participation in processing and distribution of pulses through direct marketing, farmers market or cooperative marketing to get higher share in the consumer's price. Marketing innovations like group marketing will help in improving the bargaining powers of small and marginal farmers.

The socialization model, here in this, has been christened as an alternative social process to purvey the transfer process to purvey the transfer process in a multi way channel and to a multi- dimensional projection. In the same study, the adoption, discontinuance, rejection and reinvention have been conceived as a socio-psychological polymer against a single stimulus i.e. technology exposure.

When society is getting increasingly restless owing to a series of non-compliances, conflict, comprehensive direction, mutual denial, disagreement between what we call the imposed knowledge vs inherent knowledge, exotic knowledge or exotic idea vs in- situ idea; protected need vs felt need and so on, the 'social entropy or social disorder' is expected to simmer.

Before adding new skill or useful technical knowledge, we need to study residual disorder, already created by malfunction of previous technology, and, at the same time, before adding new capacity to community capability, we need to pump out the incapability's already created by sneak of the previous technology.

That is why, the technology socialization model is an inevitable development over the transfer of technology concept, to critically analyse the sub process and sub consequence like adoption, discontinuance, rejection, and reinvention with a steamed analogy that every human mind is a complex disposition of didactic behaviour, forming what you call diodes of adoption-rejection, adoption-discontinuance, invention- reinvention, creation- culmination.

Technology socialization process has operationally been defined as the interactive summation of all possible feedback to a technology application process in terms of adoption, rejection, discontinuance and reinvention. here, this interactive summation is measured against a set of standard practices applied in pulse enterprises and the level of socialization ass measured against a "standard technology". The technology and the inputs are used as a material account of means to estimate this complex social and qualitative outcome i.e. Technology Socialization

The study has been carried out in two developed block namely Chakdah and Haringhata of Nadia District in West Bengal. Both the district and block are selected purposively due unique nature of the locations terms of technology socialization with a view to the consequence of innovation decision process viz adoption, rejection and discontinuance and market behaviour of pulse enterprise considered for present study.

Hope the study would go ahead with the help of valuable element, derived from the post- adoption drama to sake of a paradigm for researching technology socialisation vis-à-vis technology adaptation process.

In the light of the of the above discussion the researcher had delineated the following specific objectives for the present study.

## 0.1 SPECIFIC OBJECTIVES

The following specific objectives are set to intervene the present study:

1. To generate basic information on socialization of pulse crop in the study area.
2. To identify and standardize the variables, dependent and independent, impacting on both socialization and market behaviour of pulse crop in the study area.
3. To elucidate the inter and intra level interaction between dependent variables i.e. socialization and market behaviour with those of selected socio economic and ecological variables.
4. To organize participatory analysis on prospects and constraints of pulse enterprises in the study area in term of its socialization and market behaviour.
5. To delineate the micro level policy based on the empirical result on effective socialization process and market behaviour of pulse crop

2

CHAPTER

# REVIEW OF LITERATURE

In any scientific investigation a comprehensive review of relevant literature is essential. Its main function apart from, determining the work done before concerning the problem area, is to provide an insight into the methods and procedures and create a basis for interpretation of findings. As direct references on all the items are not in abundance, certain specific references along with some indirect references have been incorporated in this chapter for the purpose of meaningful use.

The review of the works has been presented under the following headings.

2.1 Socioeconomic status of pulse enterprise.

2.2 Agro economic status of pulse enterprise: a. Yield, b. Gross income.

2.3 Socialization: The most comprehensive process in the journey of technology in social ecology.

2.4 Factors related with the adoption, rejection and discontinuance of an innovation.

2.5 Market behaviour of pulse enterprise: a. Market orientation, b. Marketed surplus and marketable surplus.

## 2.1 SOCIOECONOMIC STATUS OF PULSE ENTERPRISE

| Sl.No. | Source | Year | Title | Author's | Key contents |
|---|---|---|---|---|---|
| 1 | *Indian Research Journal of Extension Education.* 2008; 8(1): **5-8** | 2008 | Extension strategies for increasing pulses production for evergreen revolution. | Burman, R.R; Singh, S.K; Lakhan Singh and Singh, A.K | The present study was conducted with a specific objective to find out the extension strategy to increase pulses production. Perception of farmers, extension personnel and research workers was |

| Sl.No. | Source | Year | Title | Author's | Key contents |
|---|---|---|---|---|---|
| | | | | | recorded to find out the constraints in different aspects of pulse based cropping system and related technologies. Among the bio-physical constraints, lack of improved varieties of pulses has been found as the most serious of the all. |
| 2 | *Indian Journal of Pulses Research.* 18(1): **60-63** | 2005 | Impact of usable technologies identified under pulse based rain fed agro ecosystem. | Singh, S.K; Burman, R.R; Chaudhary, R.G; Singh, K.K and Ansari, M.S | Found that, the economic impacts (yield increase, cost reduction, and net profit) of these technologies in Vidokhar village, Hamirpur district. |
| 3 | *Journal of Interacademicia.* 1(1): **93-96** | 1997 | Predictors of attitudinal change of pulse growers: a study with non-progressive farmers | Singh, M | This study revealed that, direct and indirect influence of various socioeconomic variables on the change in attitude. Size of family, size of holding and level of aspiration emerged as potential contributors towards the change in attitude. |

## 2.2 AGRO ECONOMIC STATUS OF PULSE ENTERPRISE- A. YIELD, B. GROSS INCOME

### A. Yield

| Sl.No. | Source | Year | Title | Author's | Key contents |
|---|---|---|---|---|---|
| 1. | *Indian Journal of Agricultural Sciences.* 84(10): **1254-1261** | 2014 | Tobit analysis of farmer to farmer diffusion of improved pulse seeds in Bundelkhand region of India. | Uma Sah; Hem Saxena; Narendra Kumar; Singh, S.K and Shripad Bhat | Finding revealed that from 2038 kg of quality seeds that was provided to 124 key farmers, about 7486 kg of produce spread as seeds among 331 farmers that covered 300.9 acres of area through farmer to farmer networks. Thus farmer to farmer networks were found to cause 3.7, 4.4 and 2.7 times enhancement in quantity of seed diffused, the area covered under the introduced varieties and farmers reached. |

| Sl.No. | Source | Year | Title | Author's | Key contents |
|---|---|---|---|---|---|
| **2** | *Indian Research Journal of Extension Education.* 14(3): **24-27** | 2014 | Impact of National Food Security Mission-Pulse on chickpea productivity in Hamirpur District of Uttar Pradesh. | Sarju Narain; Singh, S.K and Lakhan Singh | The result was found that productivity level of gram was higher in trained farmers than untrained farmers. Majority of the trained gram farmers i.e, 43.33 per cent were observed under medium productivity level i.e, 10-20 q.ha while 40 per cent achieved higher productivity level i.e, more than 20 q/ha gram yield in their field. This impact was the combination of technological package including training and critical monitory and non-monitory inputs. |
| **3** | *Communications in Soil Science and Plant Analysis.* 45(14): **1934-1948** | 2014 | Scaling up of pulse production under frontline demonstration technology transfer program in Himachal Himalayas, India. | Choudhary, A.K and Suri, V.K | It was inferred that by adopting improved pulse production technology, pulse productivity can be raised by 97-128, 39-82, 112-129, 59-65, 130-141, and 67-126% in blackgram, kidneybean, pigeonpea, cowpea, chickpea, and lentil, respectively. Improved technology package has also enhanced profitability and additional returns enhancing incremental benefit-cost ratio (1.25-7.21). |
| **4** | *Environment and Ecology.* 23(Spl-1): **20-24** | 2005 | Comparative economics of pulse crops in Rajasthan State | Ram Singh; Dalip Kumar; Suhag, K.S and Hasija, R.C | Results revealed that the highest expenses incurred in the cultivation of pulse crops were on harvesting, threshing and seed. Among the pulse crops, mung and moth beans were found to give higher returns per hectare as compared to gram. This was because of higher prices for mung and moth beans. Though the average yield of gram was observed to be higher, the net returns |

| Sl.No. | Source | Year | Title | Author's | Key contents |
|---|---|---|---|---|---|
| | | | | | per hectare of all pulse crops were low on small farms due to inefficient use of modern inputs and low yield. |
| 5 | *Current Science.* 87(7): **859-861** | 2004 | Pulse yields: feeling the pulse. | Brahmaprakash, G.P; Chandraprakash, J; Ganeshaiah, K.N and Shaanker, R.U | It presents the reasons for this inefficiency and identifies that the inputs supplied to pulses is one of the major factors limiting the pulse yields. The impact of breeding protocols, production system and policy structure for pulse cultivation are also mentioned. |
| 6 | *Journal of Research, Birsa Agricultural University.* 13(2): **211-213** | 2001 | A comparative study on the productivity level of pulses and other crops | Usha Singh | This paper reviews the farmers In the area were categorized as marginal, small, medium and large farmers according to the size of their farms. The average yield of pulse crops was the lowest among the major crops grown in the region. The small farmers had the highest average yield of pulses, followed by medium farmers, large farmers and marginal farmers. |
| 7 | *Policy Paper National Centre for Agricultural Economics.* (2): **48** pp. | 1995 | Production prospects and constraints to higher productivity of pulses in Madhya Pradesh. | Pant, S.P; Maji, C.C and Sulaiman, R.V | This paper reviews several issues surrounding the production of pulses, including the technological aspects and market incentives. It is concluded that there is a need for strengthening pulse research with specific objectives of high yield under adverse agro-ecological conditions at lower real costs of production. Marketing, processing supports and price incentives should also be ensured to make the crop as remunerative as its principal competitors.. |

| Sl.No. | Source | Year | Title | Author's | Key contents |
|---|---|---|---|---|---|
| 8 | *Indian Journal of Agricultural Economics.* 48(3): **351-356** | 1993 | Analysis of on-farm trials and level of technology on oilseeds and pulse crops in Northern Telangana zone of Andhra Pradesh | Prasad, Y.E; Rao, M.M and Vijayabhinandana, B | This paper examines: (1) comparative costs and returns of farmers' practices and on-farm trials; (2) the yield gap between farmers' practices and on-farm trials; (3) extent of adoption of improved technology; and (4) reasons for non-adoption of proven technologies. |
| 9 | *Agricultural Situation in India.* 45(2): **113-119** | 1990 | Pulses production in Bihar - an empirical analysis. | Choudhary, J.N; Singh, K.M and Singh, R.K.P | This paper revealed that: (1) examines the growth rates of area, production and yield of pigeon peas, chickpeas and lentils in 6 pulse growing zones of Bihar State; and (2) estimates factors affecting area, production and yield. The study is based on data for 1960/61-1986/87 in the case of pigeon peas and chickpeas, and 1970/71-1986/87 for lentils. |

## b. Gross income

| Sl.No. | Source | Year | Title | Author's | Key contents |
|---|---|---|---|---|---|
| 1 | *Agriculture-Update.* 6(1): **145-150** | 2011 | Impact of Krishi Vigyan Kendra trainings in adoption of biofertilizers and bio-pesticides practices by pigeonpea growers | Binkadakatti, J.S; Hanchinal, S.N and Pawar, B.B | It revealed that 47.50 per cent of trained and 12.50 per cent of untrained respondents belonged to medium adoption level category, Regarding adoption level of Rhizobium practices, 43.75 and 11.25 per cent of trained and untrained farmers were of high adoption category, Majority of 47.50 and 90.00 per cent of trained and untrained farmers were of low adoption category with respect to adoption level of phosphorus solubilizing bacteria practices, Regarding the adoption level of utility of NSKE practices, half of the trained (50.00%) respondents and only 15.00% of untrained respondent were of medium adoption category. |

| Sl.No. | Source | Year | Title | Author's | Key contents |
|---|---|---|---|---|---|
| 2 | *Agriculture Update.* 4(3/4): **262-265** | 2009 | Economic analysis of pulses on medium farms in Marathwada region of Maharashtra | Asmatoddin, M; Jawale, S.V and Perke D.S | The results revealed that, in case of soybean per hectare cost of cultivation i.e. cost 'C' was Rs. 11355.60 and net profit was Rs. 4761.86. In case of pigeon pea per hectare cost of cultivation i.e. cost 'C' was Rs.12632.42 and net profit was Rs.2012.16. In green gram, per hectare cost of cultivation i.e. cost 'C' was Rs.8935.48 and net profit was Rs.3292.80. |
| 3 | *Annals of Agri Bio Research.* 12(1): **63-64** | 2007 | Economics of production and marketing of urd pulse crop in Shivalik foothills of Haryana. | Singh, U.B; Satpal Baloda; Sangwan, N; Gurnam Singh and Sandeep Antil | The cost of cultivation per ha was Rs.5916 and 6274 in the upper and lower project areas, respectively. The profitable net return of Rs.1325/ha was obtained in the lower project area. The crop had much higher potential of yield if disease resistance and high yielding varieties were developed. |
| 4 | *Indian Journal of Pulses Research.* 18(1): **60-63** | 2005 | Impact of usable technologies identified under pulse based rainfed agro ecosystem. | Singh, S.K; Burman, R.R; Chaudhary, R.G; Singh, K.K and Ansari, M.S | Found that, economic impacts (yield increase, cost reduction, and net profit) of these technologies in Vidokhar village, Hamirpur district. |
| 5 | *Research Study Agro Economics Research Centre, Sardar Patel University.* (110): **58** pp. | 1999 | Economics of pulses production and identification of constraints in raising their production in Gujarat. | Patel, V.G | The constraints identified in the cultivation of pulses are that most of farmers lacked modern farm inputs with marginal variations in land owned and operated. Poor irrigation and the cultivation of a long duration cotton crop led to low cropping intensity, and a change in cropping pattern in favor of food crops. The productivity of pulses was low and its cost of cultivation was high as compared to other crops. The study provides several suggestions to increase pulse production in the state. |
| 6 | *Agricultural Situation in India.* 44(4): **283-284** | 1989 | Prospects of changing cropping pattern in favour of oilseeds and pulses production in Assam | Saikia, T.N | The study was based largely on data from the Cost of Cultivation Scheme for the years 1971/72, 1980/81 and 1984/85. Due to physiological and other factors a major shift in the cropping pattern in favor of oilseeds and pulse crops may not be possible. In certain suitable areas, an area approach planning method is needed to increase production. |

## 2.3 SOCIALIZATION: THE MOST COMPREHENSIVE PROCESS IN THE JOURNEY OF TECHNOLOGY IN SOCIAL ECOLOGY

| Sl.No. | Source | Year | Title | Author's | Key contents |
|---|---|---|---|---|---|
| 1 | *Environment and Ecology.* 31(4A): 1890-1894 | 2013 | Socialization of rubber enterprise: a multifactorial study on livelihood security | Biswas, A; Acharya, S.K; Riyang, K and Manashi Saha | Found that, resource entrepreneurship has been found to be important causal factor to characterize the dictum and direction of socialization of plantation crop. Contemplating these factors are crucial in modeling socialization in any given agro-socio-eco system at both micro and macro level. |
| 2 | *Ecological Economics.* **68**(10): 2721-2728. | 2009 | Estimating the social cost of pesticide use: an assessment from acute poisoning in Brazil. | Soares, W.L. and M.F.de.S. Porto | Found the “invisible” or social, environmental and health costs which end up being socialized with the farmer, in general, having no incentives to recognize and internalize them. |
| 3 | *Agrarforschung.* **16**(6): 186-191. | 2009 | Rural innovation networks: from knowledge to development. | Hartwich, F. | The social connectivity between the farmers and also with various in the agro-industry and in research and development is an important parameter in determining the innovative behaviour of farmers. The rule is: he who networks is more innovative. |
| 4 | *Population, Space and Place.* **15**(3): 253-266. | 2009 | Whose socialization? Exploring the social interaction between migrants and communities-of-place in rural areas. | Vergunst, P.J.B. | Found that the people belonging to the communities-of-place in the Dutch study do not hold rigidly to some elements of their institutions while they hold on more rigidly to others. The Scottish study showed that farmer-employers even prefer the work ethic of migrants and at the same time there has been some evidence that the work ethic of domestic workers evolves in the direction of that of migrants. |
| 5 | *Revue d'-Etudes en Agriculture et Environnement.* **88**: 71-94. | 2008 | Women starting up in agriculture: from gender socialization to training. | Rieu, A. and S.Dahache | Found that the agricultural profession has gradually opened up to women, the reproduction strategies of families through socialization, male favoritism in inheritance practices, etc., represent considerable barriers to their professional choice. |

| Sl.No. | Source | Year | Title | Author's | Key contents |
|---|---|---|---|---|---|
| 6 | *Revue Suisse d'Agriculture.* **40**(3): 120-122. | 2008 | Farm succession: interest and motivation of the coming generation. | Rossier, R. | Found that, this is partially the result of a gender oriented and farm their specific socialization. |
| 7 | *Technology in Society.* **28**: 393-406. | 2006 | The importance of social context influences on new farm technology sustainability: community and sub-community characteristics in Jamaica. | Moxley, R.L. and K.B.Lang | In a study of Jamaican farmers, results suggest that the farmer's community and social context are more important than the farm and personal characteristics in influencing the long term sustainability of a farming innovation |
| 8 | *Tourism ecreation-Research.* **24**(1): 82-85. | 1999 | Senior women's perception of leisure in India. | Prakash, I.J. | Influenced by their early socialization, present health and economic status, they structure their time in and around their household. |
| 9 | *Medecine et Nutrition.* **30**(4): 171-177. | 1994 | Psychology and nutrition: study of the process of food socialization. | Watiez, M. | Found that the, it is a process by which children develop taste, knowledge, opinions and food related behaviour so as to adapt to the eating habits of the sociocultural group to which they belong. |

## 2.4 FACTORS RELATED WITH THE ADOPTION, REJECTION AND DISCONTINUANCE OF AN INNOVATION.

| Sl.No. | Source | Year | Title | Authors | key contents |
|---|---|---|---|---|---|
| 1. | *Journal of the Indonesian Tropical Animal Agriculture.* 39(1): 58-64 | 2014 | Path analysis of recording system innovation factors affecting adoption of goat farmers. | Okkyla, S; Dwijatmiko, S and Kurnianto, E | The result of path analysis showed that the influence of motivation, innovation characteristics and information source on adoption were 0.168; 0.720 and 0.09, respectively. Innovation characteristics were the greatest effect on adoption. In conclusion, by improving innovation characteristics of respondent through motivation and information source may significantly increase the adoption of recording system in goat farmers. |

| Sl.No. | Source | Year | Title | Authors | key contents |
|---|---|---|---|---|---|
| 2 | *Indian Research Journal of Extension Education.* 14(3): 125-127 | 2014 | Knowledge and adoption level of IPM practices among tomato growers in Indore District (M.P.). | Surat Singh and Sarju Narain | The study finds out large gap between knowledge and adoption in respect to IPM practices. So, it is an argent need of skill oriented training to tomato growers regarding IPM practices for higher and safer tomato production. |
| 3 | *Environment and Ecology.*; 32(4): 1342-1346 | 2014 | Adoption of modern agricultural technologies in Bihar: a farm level study. | Singh, K.M; Singh, R.K.P and Abhay Kumar | Analysis revealed that KVKs covered less than one per cent of villages under their jurisdiction, therefore govt departments were mainly responsible for implementation of agricultural development. Overall, only 22.8% farmers used modern seeds however, it was high at 46.67% for medium farmers. About 50% large farmers adopted scientific method of production of horticultural crops |
| 4 | *Agrociencia Montevideo.*; 15(2): 158-163 | 2011 | Type of subjectivity associated with low technology adoption in fruit culture. | Fuster, R; Felix, A; Hegedus Hetze, P de and Gravina Tejera, M.V | The study revealed problems from the extension perspective: (i) in the conceptual level, because the family and neighbors of the producer are not taken into account, although they play a big role in how a decision is made by the producer, and (ii) in the operative level, the instruments used by extension (for example, inappropriate use of credit). |
| 6 | *Indian Res. J Ext. Edu.* **9** (2 ):39-45. | 2009 | A Logit Analysis of Bt Cotton Adoption and Assessment of Farmers' Training Need. | Padaria *et al.* | Found significant influence of size of holding, capital base, extension contact, innovativeness, achievement motivation, and perception about Bt cotton on adoption decision of the farmers for Bt cotton, whereas in contrary to a priori expectation, information source pluralism, mass media exposure, social participation and education were not found to have a significant influence. |

| Sl.No. | Source | Year | Title | Authors | key contents |
|---|---|---|---|---|---|
| **7** | *Indian Res. J. Ext. Edu.* **9** (1):54-57. | 2009 | Communication Pattern in Drylands of Uttar Pradesh. | Pal *et al.* | The study revealed that the access of different cosmopolite sources was low as compared to interpersonal localite sources. Possession of land holdings and use of information sources are positively and significantly correlated, except in case of radio. It was found that educational level increases, the use of information sources. |
| **8** | *Indian Res. J Ext. Edu.* **9** (2):80-84. | 2009 | Adoption of Improved Dairy Cattle Management Practices under Vidarbha Development Programme Package. | Khode *et al.* | Found that the education and socioeconomic status were found highly significant. Whereas land size, total annual income, dairy herd size, daily milk production, daily milk sale, milk production from purchased dairy cattle social participation, utilization of communication sources, knowledge level, attitude towards dairy farming, economic motivation and training on dairy farming were significantly correlated with adoption of improved dairy cattle management practices. |
| **9** | *World Applied Sciences Journal.* **6** (5): 644-651. | 2009 | Analysis of Factors Affecting Adoption of Sustainable Soil Conservation Practices among Wheat Growers. | Rezvanfar *et al.* | The results of regression analysis show that level of knowledge could explain 83.5 per cent of the variation in the adoption level of sustainable soil conservation practices. |
| **10** | *Environment for Development Discussion Paper Resources for the Future RFF-* (09-15): iii + 17 pp. | 2009 | Farmers' preferences for crop variety traits: lessons for on-farm conservation and technology adoption. | Asrat *et al.* | Found that the household resource endowments (particularly land holdings and livestock assets), years of farming experience, and contact with extension services are the major factors causing household heterogeneity of crop variety preferences. |

| Sl.No. | Source | Year | Title | Authors | key contents |
|---|---|---|---|---|---|
| 11 | *Agriculture Update.* **4**(1/2): 59-61. | 2009 | Technological gap in adoption of improved mango production technology. | Jadav, N.B. and M.M Solanki,. | Found that the almost all of the independent variables except for age and size of family were negatively correlated with the extent of the technological gap in the adoption of improved mango production technology. |
| 12 | *Australian Journal of Basic and Applied Sciences.* **2**(4): 823-834. | 2008 | Factors influencing adoption of improved maize varieties in Nepal: a case study of Chitwan district. | Paudel, P. and A. Matsuoka | Found that the adoption of improved maize varieties is significantly influenced by winter maize cultivation, education of the household head, lowland area, upland area as well as access to credit and extension services. |
| 13 | *Journal of Extension Systems.***24**(1): 76-93. | 2008 | Factors that Influence the Adoption of Cocoa Integrated Pest Management by Farmer Field School Graduates in Cameroon. | Lapbim *et al.* | The results show that the factors that positively influence Farmers decision to apply the IPM technological package are the level of education, family size, membership to a farmer organization and previous knowledge of improved spraying practices. Factor found to negatively affect the application of IPM technological package was the size of the coco farm. |
| 14 | *Scientia Agricultura Sinica.* **41**(7): 1963-1968. | 2008 | Analysis of farmers' acceptance and response to agricultural technologies in main grain production district - a case from Yushu County in Jilin. | QiFeng *et al.* | Found that the 38.4 per cent of the farmers accepted new technologies mainly through agricultural extension, 64.9 per cent of the farmers learned about new technologies from neighbouring farmers, and 89.4 per cent of the farm households accepted agricultural technology extension, but 50 per cent of the households had no access to extension services |
| 15 | *Indian Res. J. Ext. Edu.* **8** (2&3) : 55-56. | 2008 | Effect of Motivation and Knowledge on Decision Making and Adoption of Dairy Practices. | Gautam *et al.* | Found that the recognition motive, security motive, scientific motive, affiliation motive and total score of motivation found to be significantly correlated with adoption. Motivation was |

| Sl.No. | Source | Year | Title | Authors | key contents |
|---|---|---|---|---|---|
| | | | | | significantly correlated with decision making component and adoption of dairy practices. |
| **16** | *Indian Res. J. Ext. Edu.* **8** (2&3) : 65-69 | 2008 | Knowledge Level of Chickpea Growers about Chickpea Production Technology. | Shakya *et al.* | The socio-economic, communication and psychological factors had significant positive relationship with knowledge level of chickpea growers except age, land holding, and farm mechanization. |
| **17** | *Indian Res. J. Ext. Edu.* **8** (2&3) : 89-91. | 2008 | Awareness, Attitude and Training Needs of Farmers about Recommended Practices in Watershed Development Programme. | Rai, D.P. and Singh ,K. | Found that the good association exists between three categories of socio-economic status and training needs |
| **18** | *Indian Res. J. Ext. Edu.* **7** (1) : 41-43. | 2007 | Knowledge and Adoption of Agricultural Technologies in Marathwada. | Deshmukh *et al.* | Found that the majority of the respondents (97.92 per cent) belong to the low level of knowledge and 81.94 per cent were falling under low level of adoption. |
| **19** | *Journal of Extension Systems.* **23**(1) : 70-80. | 2007 | Appropriateness of E-learning Based Information Technology to Improve the Productivity of Crops | Ommani A.R. and C. Mohammad | Found the significant relationship between the IT skill and knowledge, income, social participation, the extent of information-seeking motivation, level of job satisfaction, and level of education with perceptions of agricultural extension agents regarding the appropriateness of e-learning-based Information Technology. |
| **20** | *Journal of extension* .**45** (6) Feature Articles , 6FEA6 | 2007 | Organic Farmers' Need for and Attitude Towards Extension. | Agunga, R. and C. Igodan, | Found that the sustainable agriculture farmers want to learn from each other or share their experiences with one another. |
| **21** | *Journal of Cotton Research and Development.* **21**(2): 230-234. | 2007 | IPM in cotton - adoption and constraints. | Yadav *et al* | Found positive and significant correlation related to IPM adoption namely, land holding, socio-economic status, land under cotton crop, information seeking behaviour, extension participation, risk |

| Sl.No. | Source | Year | Title | Authors | key contents |
|---|---|---|---|---|---|
| | | | | | orientation, economic motivation, management orientation and innovativeness. |
| **22** | *Journal of Extension Systems* .**22**(2): 01-13. | 2006 | Socio-Economic Impact of Cocoa Crop and Pest Management Knowledge through Farmer Field School (FFS) Approach in Cameroon,. | Tchouamo *et al.* | Found that the IPM practices lowered overall costs of production by 11 per cent relative to previous practices. |
| **23** | *Journal of Extension Systems* .**19**(1):45-53. | 2003 | Analysis of Mass Media Use for Agricultural Information by Farmers in Nigeria. | Ajayi, M.T. | Found significant relationships between the use of mass media and farmers' level of education, annual income and membership of cooperative society. |
| **24** | *Journal of Science and Technology of Agriculture and Natural Resources*. **6**(4): 53-65. | 2003 | Economic factors effective in adopting an integrated campaign in rice stem borer control in Isfahan. | Pezeshki, R.G. and M. Masaeli | Found that the adoption was significantly related to area under cultivation, degree of family cooperation in agricultural activities, access to farm inputs, access to financial resources, and yield of rice per hectare. |
| **25** | *J. Manage. Dev.* **19** (9): 783–793. | 2000 | Knowledge value chain. | Lee, C. and J.Yang | Knowledge management is about people and the processes they use to gather data, share information, transform knowledge, teach, learn, apply and innovate. |
| **26** | *Savings and Development.* **20**(4): 431-445. | 1996 | "Adoption of formal agricultural credit by opinion leaders and other farmers in differentially developed villages of Bangladesh". | Charmala, S. and S.M.A. Hossain | Have identified farm and technology specific factors, institutional, policy variables, and environmental factors to explain the patterns and intensity of adoption. |
| **27** | *Quarterly Journal of International Agriculture.* **35** (4): 325-337. | 1996 | "Factors affecting the adoption of wheat production technology in Sudan". | Abdelmagid, S.A and F.K .Hassan, | Have identified farm and technology specific factors, institutional, policy variables, and environmental factors to explain the patterns and intensity of adoption. |

| Sl.No. | Source | Year | Title | Authors | key contents |
|---|---|---|---|---|---|
| **28** | *Journal of research and ANGRAU 24* (1-2): 21 – 25. | 1996 | "Adoption of rice production technology by tribal farmers". | Rao, P.P and V.G.K Rao | Found a positive and significant association between age, farming experience, training received, socio-economic status, cropping intensity, aspiration, economic motivation, innovativeness, information source utilization, information source, agent credibility and adoption |
| **29** | *Journal of Extension System.* **10** (1): 46–58. | 1994 | Discriminant analysis of small holder farmer adoption potential and the prediction of extension cost in Nigeria: a comparative enterprise perspective. | Arene, C.J | Reported a positive and significant relationship between family size and adoption. |
| **30** | *West Africa Agricultural Economics. 13(1995) 1-9* | 1995 | Farmers perceptions and adoption of new agricultural technology: evidence from analysis in Burkina Faso and Guinea. | Adesina, A.A and Baidu J Forson | Found that farmers perceptions affect the adoption of improved varieties of sorghum and mangrove rice in Burkina Faso and Guinea respectively. |
| **31** | *Farm information source utilization, adoption and transfer behaviour of farmers; Rural India.* **57**(7): 153-154. | 1994 | Farm information source utilization, adoption and transfer behaviour of farmers; Rural India. | Bhople, R. S . and S .M.Thakare | The adoption behaviour of the dairy farmers depends on education, knowledge, attitude, risk orientation and innovation proneness |
| **32** | *Indian Journal of Animal Sciences.* **64**(5): 501-507 | 1994 | A comparative study of adoption of dairy farm technologies between non-members and members of dairy co-operative villages. | Kunzru, O .N . and Tripathi ,H. | The adoption behaviour of the dairy farmers depends on education, knowledge, attitude, risk orientation and innovation proneness |

| Sl.No. | Source | Year | Title | Authors | key contents |
|---|---|---|---|---|---|
| 33 | *Journal of Rural Development and Administration.* xxv (3): 111 – 113. | 1993 | Farmers characteristics affecting adoption of agricultural innovations | Abdul *et al.* | Reported a significant relationship between landholdings (farm size) and adoption and did not establish any relationship between education and adoption. Education, size of holdings and cosmopoliteness accounted for significant variation in communication behaviour of farmers. |
| 34 | *Journal of Agricultural Economics Research.* 42(3): 20-26 | 1990 | Partial adoption of divisible technologies in agriculture | Szmedra, P.I; Wetzstein, M.E and Mc Clendon, R.W | Results indicate that the degree of new technology adoption may depend on the extent of the risk. For example, strongly risk-averse producers who use dryland technology may only partially adopt IPM. And producers who irrigate to significantly decrease variation in yield and returns may also only partially adopt IPM. |

## 2.5 SOCIALIZATION OF PULSE ENTERPRISE: ADOPTION, KNOWLEDGE LEVEL, ATTITUDE AND DISCONTINUANCE OF AN INNOVATION IN PULSE ENTERPRISE.

| Sl.No. | Source | Year | Title | Author's | Key contents |
|---|---|---|---|---|---|
| 1 | *Journal of Agricultural Sciences Sri Lanka.* 9(2): 78-87 | 2014 | Factors affecting the discontinuation of protected agriculture enterprises in Sri Lanka | Wijerathna, M; Weerakkody, W.A.P and Kirindigoda, S | Found that, the main reasons for discontinuing protected agriculture enterprises were socio-economic factors, namely; lack of transport facilities, problems in the marketing, marginal monitory returns, dependency on external support schemes on investments, part time engagement, lack of family support and scale of production (less than 2000 sq ft.). |
| 2 | *Indian Research Journal of Extension Education.* 14(2): 20-24 | 2014 | Assessment of gaps in pulse production in Hamirpur district of Himachal Pradesh. | Sanjay Kumar; Ravinder Singh and Akhilesh Singh | There was a wide yield gap between the potential and demonstration yields in both pulse crops mainly due to technology and extension gaps. The frontline |

| Sl.No. | Source | Year | Title | Author's | Key contents |
|---|---|---|---|---|---|
| | | | | | demonstrations on black gram crop indicated that per cent increase in yield over farmers practice ranged from 31.1 to 76.6 per cent where as in case of chickpea the range was 39.4 to 63.6 per cent over five years. |
| 3 | *Indian Research Journal of Extension Education.;* 14(3): 24-27 | 2014 | Impact of National Food Security Mission-Pulse on chickpea productivity in Hamirpur District of Uttar Pradesh. | Sarju Narain; Singh, S.K and Lakhan Singh | The result was found that productivity level of gram was higher in trained farmers than untrained farmers. Majority of the trained gram farmers i.e, 43.33 per cent were observed under medium productivity level i.e, 10-20 q.ha while 40 per cent achieved higher productivity level i.e, more than 20 q/ha gram yield in their field. |
| 4 | *Bangladesh Journal of Agricultural Research;* 39(1): 113-125 | 2014 | Adoption of mungbean technologies and technical efficiency of mungbean (Vigna radiata) farmers in selected areas of Bangladesh | Haque, M.A; Miah, M.A.M; Ali, A.M and Luna, A.N | The study assessed the farm level adoption of mungbean technologies, technical efficiency of mungbean growers, and find out constraints to its higher production. |
| 5 | *Indian Journal of Soil Conservation.* 41(1): 88-97 | 2013 | Technological and extension yield gaps in pulse crops in Mandi district of Himachal Pradesh, India. | Choudhary, A.K | Extension yield gaps varied to the extent of 368-492 kg ha-1 in black gram (Vigna mungo), 220-417 kg ha-1 in kidney bean (Phaseolus vulgaris), 477-563 kg ha-1 in pigeonpea (Cajanus cajan), 372-494 kg ha-1 in cowpea (Vigna sinensis), 225-601 kg ha-1 in chickpea (Cicer arietinum) and 253-510 kg ha-1 in lentil (Lens culinaris) during study period. |
| 6 | *International Journal of Bio resource and Stress Management.* 4(2): 293-297 | 2013 | Strategies to overcome pulse production constraints in West Bengal, India. | Ray, M; Ghosh, P.K and Kajal Sengupta | It means, the state has to produce 10 times its present production to meet the demand. In the last 15 years, area of pulses declined due to cropping system shift to vegetables and boro rice with introduction of irrigation facilities. |

| Sl.No. | Source | Year | Title | Author's | Key contents |
|---|---|---|---|---|---|
| 7 | *Bangladesh Journal of Agricultural Research.* 37(4): 593-606 | 2012 | Adoption and profitability of BARI lentil varieties in some selected areas of Bangladesh. | Rahman, M.S; Hossain, M.A; Sarker, M.J.U and Bakr, M.A | The study revealed that 98% of the total lentil cultivated areas were occupied by BARI lentil varieties in the study areas. The average level of adoption of BARI Masur-3, BARI Masur-4, BARI Masur-5 and BARI Masur-6 were 49%, 47%, 1% and 1%, respectively at farm level. |
| 8 | *Legume Research* 34(3): 184-189 | 2011 | Knowledge and adoption level of Rhizobium culture among pulse cultivators of Orissa | Nath, S.K and Mohapatra, B.K | It was revealed that only 4 per cent of the respondents had adopted it, whereas 41 per cent were yet to hear of it. Television was used as the source of information by the highest numbers of farmers (34%) followed by radio (28%). Nearly half of the responding farmers did not possess the knowledge of the recommended dose. |
| 9 | *Outlook On Agriculture.* 40(3): 245-250 | 2011 | Adoption, impact and discontinuance of integrated pest management technologies for pigeon pea in South India. | Rao, C.A.R; Rao, M.S; Srinivas, K; Patibanda, A.K and Sudhakar, C | The adoption and impact of integrated pest management (IPM) were examined for pigeon pea cultivation in Andhra Pradesh, India. Summer ploughing and spraying of Neem Seed Kernel Extract (NSKE) were the most regularly adopted components of IPM for pigeon pea. |
| 10 | *Legume Research.* 34(1): 8-13 | 2011 | Spread and acceptance of recommended production technologies in green gram and black gram: a comparative analysis. | Arunachalam, R | The pure crop respondents (Black gram) have fairly adopted the technologies viz., right time of sowing, optimum seed rate, recommended fertilizer, regular weeding, control of pod borer and powdery mildew and harvesting at the right time. |
| 11 | *Legume Research.* 34(3): 184-189 | 2011 | Knowledge and adoption level of Rhizobium culture among pulse cultivators of Orissa. | Nath, S.K and Mohapatra, B.K | It was revealed that only 4 per cent of the respondents had adopted it, whereas 41 per cent were yet to hear of it. Television was used as the source of information by the highest numbers of farmers (34%) followed by radio (28%). Nearly half of the responding farmers did not possess the knowledge of the recommended dose. |

| Sl.No. | Source | Year | Title | Author's | Key contents |
|---|---|---|---|---|---|
| 12 | *Environment and Ecology.* 29(1A): 243-246 | 2011 | Participatory approaches for productivity enhancement in pigeon pea through improved production technologies. | Kantharaju, V; Basavaraja; Teggelli, R.G and Girish, G | The improved technologies recorded a mean yield of 13.54 q/ha which was 18.69% higher than that obtained with farmers practice yield (11.10 q/ha) besides the higher mean net income of Rs 22,876/ha with a B:C ratio of 2.68 when compared to farmers practice (Rs 16,177/ha and 2.12). |
| 13 | *Agriculture Update.* 5(3/4): 352-355 | 2010 | Adoption behaviour of farmers in chickpea production technology. | Thoke, N.J and Surya Gunjal | The study reveals that most of the respondents were from middle age group i.e. between 26 to 45 years, received education up to higher Secondary, size of land holding between 4.01 to 7.00 acres. Majority (54.46 per cent) of chickpea growers were having farming experience between 9 years to 17 years, having medium social participation and annual income between Rs. 75,551 to Rs. 1,50,765. Most of the respondents (60.72 per cent) were having their land at 2-3 places. |
| 14 | *Agriculture Update.* 5(1/2): 52-53 | 2010 | Front line demonstration: an effective technology for increasing productivity of chickpea in Ratlam district. | Subhash Katare and Shrivastava, M.K | The productivity of pulse crops continues to be quite low due to technological gap in adoption pulses of technology. The yield of pulses can be increased by demonstrating their cultivation technologies at farmer's field under the supervision of scientist working in operational area. Keeping the above fact front line demonstration were under taken by the Krishi Vigyan Kendra Jaora. |
| 15 | *Electronic Journal of Plant Breeding.* 1(4): 1099-1106 | 2010 | Role of classical breeding in improvement of pulse crops. | Nadarajan, N and Sanjeev Gupta | Found that, the total of 513 cultivars of different pulse crops including chickpea, pigeonpea, mungbean, urdbean, lentil, fieldpea and rajmash were developed in India itself during last three decades. Adoption of high yielding varieties in early 1980's increased the average productivity of the country by 34% now. |

| Sl.No. | Source | Year | Title | Author's | Key contents |
|---|---|---|---|---|---|
| **16** | *Indian Research Journal of Extension Education.* 10(1): 99-104 | 2010 | Gap in adoption of improved pulse production technologies in Uttar Pradesh | Burman, R.R; Singh, S.K and Singh, A.K | Found that, the gaps in technology adoption in major pulse crops had been identified both in rainfed and irrigated cropping. The overall gap in adoption of technologies was larger in the rainfed situation than in the irrigated situation. |
| **17** | *Legume Research.* 33(2): 102-107 | 2010 | Cognition and adoption of pulse production technologies by dry land farmers of Tamilnadu. | Ramasubramanian, M; Seeralan, S; Sekar, V and Thomas, M I | From the study, it is found that social participation, decision making pattern, economic motivation, information seeking behaviour are the essential factors which critically influence knowledge and adoption level of pulse growers and subsequently the production of pulses. |
| **18** | *Quarterly Review of Biology.*84(1): 29-50 | 2009 | Reconsidering domestication of legumes versus cereals in the ancient Near East. | Abbo, S; Saranga, Y; Peleg, Z; Kerem, Z; Lev Yadun, S and Gopher, A | In this paper discuss, from both biological and cultural perspectives, the ancient human-plant liaison that gave rise to Near Eastern agriculture. We explain the biological aspects of Near Eastern plant domestication by a comparative analysis of legume vs. cereal crop evolution. |
| **19** | *Indian Research Journal of Extension Education.* 8(1): 5-8 | 2008 | Extension strategies for increasing pulses production for evergreen revolution. | Burman, R.R; Singh, S.K; Lakhan Singh and Singh, A.K | Find out the constraints in different aspects of pulse based cropping system and related technologies. Among the bio-physical constraints, lack of improved varieties of pulses has been found as the most serious of the all. |
| **20** | *Journal of Food Legumes.;* 21(2): 140-142 | 2008 | Sustainability of pulse-based cropping systems in Uttar Pradesh. | Burman, R.R; Singh, S.K and Singh, A.K | It was revealed that the Sustainability Yield Index (SYI) value of this system was very low (0.51). Although the yield of chickpea grown in fallow was higher than chickpea grown after urd bean, the SYI value of fallow - chickpea was lower (0.66) than that of urdbean - chickpea (0.67). |

| Sl.No. | Source | Year | Title | Author's | Key contents |
|---|---|---|---|---|---|
| 21 | *Economic Affairs Calcutta.* 47(2): 100-107 | 2002 | Adoption of HYV chickpea and the profitability of chickpea cultivation in the High Barind Tract of Bangladesh. | Islam, Q.M.S; Karim, M.R; Ali, M.O and Rahman, M.M | It was revealed that 58% of the farmers adopted HYV chickpea, and the area planted to HYV chickpea covered 60% of chickpea cultivated land. On average, chickpea farmers used 53 man days, 50 kg of seeds, and 152 kg of manures per hectare. The average chickpea yield was 1131 kg/ha |
| 22 | *Karnataka Journal of Agricultural Sciences.* 10(1): 177-182 | 1997 | Impact of adoption of recommended practices on yield of pulse crops. | Dolli, S.S and Swamy, B.S | The study assessed significant difference in the mean yields of adopters and non-adopters of variety and seed rate practices for all three pulses; non-significant difference was observed with respect to spacing and rhizobium treatment practices. |
| 23 | *Indian Journal of Agricultural Economics.;* 48(3): 351-356 | 1993 | Analysis of on-farm trials and level of technology on oilseeds and pulse crops in Northern Telangana zone of Andhra Pradesh. | Prasad, Y.E; Rao, M.M and Vijayabhinandana, B | The findings show the extension gap to be greater for groundnuts, sesame and green gram, in addition to the low level of adoption of technology. The area under sunflowers is increasing and marketing is becoming a problem; due to lack of processing capacity in the region, cultivators are receiving low prices. |

## 2.6 MARKET BEHAVIOUR OF PULSE ENTERPRISE:

### a. Market orientation

| Sl.No. | Source | Year | Title | Author's | Key contents |
|---|---|---|---|---|---|
| 1 | *International Research Journal of Agricultural Economics and Statistics.* 4(1): 115-116 | 2013 | Marketing behaviour of pulse growers of Amravati district of Maharashtra | Kad, R.G; Deshmukh, A.N; Mokhale, S.U; Angaitkar, A.G | Found that, majority (73.33%) of farmers had medium level of marketing behaviour. Among the selected characteristics, education, Cosmo politeness, extension contact, market orientation and mass media availability were positive and significantly related with level of marketing behaviour and age and farming experience were negative and significant with marketing behaviour whereas land holding and annual income were non-significant with level of marketing behaviour. |

| Sl.No. | Source | Year | Title | Author's | Key contents |
|---|---|---|---|---|---|
| 2 | *Chickpea and pigeon pea economies in Asia:* facts trends and outlook. vi + 68 pp | 2010 | Chickpea and pigeon pea economies in Asia: facts, trends and outlook. | Rao, P.P; Birthal, P.S; Bhagavatula, S and Bantilan, M.C.S | This report provides a factual assessment of consumption, production and marketing of pulses in Asia, and explores future prospects for the pulses sector by focusing on chickpea and pigeonpea, the two most important pulse crops grown in the continent. Chickpea and pigeonpea are important grain legumes and play a significant role in the food and nutrition security of the poor in developing countries of Asia. |
| 3 | *Proceedings of the Annual Research Conference Agricultural Sciences, Yangon, Myanmar*, 10-12- 295-317 | 2004 | Demand and supply analysis of Myanmar's agricultural export: the case of pulses | Thanda-Kyi and von Oppen | This study determines the prospects for Myanmar's pulses export expansion strategy. Specifically, it (a) analyses the historical determinants of supply and demand for pulses; (b) estimates the relative price and income elasticities of demand and the price elasticity of supply for pulse export; (c) evaluates the relative price and income elasticities of actual exports of pulses; and (d) draws policy recommendations for improving Myanmar's export performance in the world market. |
| 4 | *Palawija News.* 14(1): 12-19 | 1997 | Production, consumption and marketing of pulses in Myanmar. | Hla Kyi | Found that, major factors affecting the production of pulses (policy, area expansion, low yielding varieties, agro-technology and inputs); pulse consumption; marketing (market structure, price, marketing cost, and margin and export market). Recommendations are presented within the conclusion to improve the pulse trade. |
| 5 | *Policy Paper National Centre for Agricultural Economics.* (2): 48 pp. | 1995 | Production prospects and constraints to higher productivity of pulses in Madhya Pradesh | Pant,S.P; Maji, C.C and Sulaiman, R.V | The paper reviews several issues surrounding the production of pulses, including the technological aspects and market incentives. Marketing, processing supports and price incentives should also be ensured to make the crop as remunerative as its principal competitors. |

| Sl.No. | Source | Year | Title | Author's | Key contents |
|---|---|---|---|---|---|
| 6 | *Economic and Political Weekly.* 29(26): A-89-A-100 | 1994 | Chickpea competitiveness in India | Kelley, T.G and Rao, P.P | The paper estimates national and regional trends in chickpea crop area, identifies factors underlying these trends, and assesses the demand and regional supply prospects to the year 2000. Where possible, data are disaggregated by cultivation under irrigated and rain fed conditions. An analysis of chickpea's performance provides the basis for assessing regional supply prospects for chickpea in India in the 1990s and beyond. Per caput consumption of chickpea and other major pulses are also examined. |
| 7 | *Economic Affairs.* 36(3): 145-154 | 1991 | Pulse marketing structure and its constraints in Bangladesh. | Hussain, M.S; Sikder, F.S and Elias, S.M | Found that. the marketing structure for pulses in the country drawing on data from 6 districts. The performance of different intermediaries in the marketing system is also assessed and suggestions made to overcome present constraints to marketing. Farmers are found to be price responsive, indicating the potential success of a national price policy which would give producers incentives to market more pulses |

## b. Marketed surplus and Marketable surplus

| Sl.No. | Source | Year | Title | Author's | Key contents |
|---|---|---|---|---|---|
| 1 | *Legume Research.* 34(4): 267-272 | 2011 | Analysis of marketable and marketed surplus of red gram in Vadodara district of Gujarat. | Amruta Borate; Zala, Y.C and Darji, V.B | Found that, the that marketable surplus was positively and significantly related with cropped area and average productivity in all the four categories of farms. It was negatively related with family size and quantity retained for wages in kind indicating inverse relationship between extent of marketable surplus and these factors. Further, in case of marketed surplus, the examination of individual coefficients revealed that marketed surplus was positively and significantly related with |

| Sl.No. | Source | Year | Title | Author's | Key contents |
|---|---|---|---|---|---|
| | | | | | total production, current prices and financial obligation for sample as a whole while the family size showed negative sign indicating inverse relationship of marketed surplus with family size in sample farms. |
| 2 | *Agricultural Economics Research Review.* 24(2): 267-273 | 2011 | Marketing efficiency of green peas under different supply chains in Punjab. | Sidhu, R.S; Sidhu, M.S and Singh, J.M | Found that, the marketing of green peas has been studied by three supply chains, viz. I: Producer -> wholesaler (through commission agent) -> retailer -> consumer; II: Producer -> retailer (through commission agent) -> consumer; III: Producer -> consumer. |
| 3 | *Agricultural Economics.* 25(1): 103-118 | 2001 | Modeling the impact of technological change on nutrition and marketed surplus. | Ahmed, M.M; Preckel, P.V; Baker, T.G; Lopez Pereira and M | Found that, the expected direct utility maximization model allows the estimation of the effects of new technology on consumption and marketed surplus in situations where marginal values of products vary by state of nature and are affected by consumption and production choices. |
| 4 | *Agricultural Marketing.* 40(3): 28-30 | 1997 | Marketed surplus and price spread of green gram: a case study in Parbhani district of Marathwada region of Maharashtra. | Pawar, N.D | Results are presented of a study which examined: the marketable and marketed surplus of green gram (moong) by size of holding (small, medium, large); major marketing channels; and price spread. Data were collected from a sample of 79 cultivators from three villages in Parbhani district, Maharashtra, India year not given. |
| 5 | *Bihar Journal of Agricultural Marketing.* 2(2): 153-163 | 1994 | Regional variations in food grains, marketable and marketed surplus in Uttar Pradesh. | Bajpai, B.K | Found that, the factors responsible for regional variations in marketable and marketed surplus of food grains in Uttar Pradesh, India. A multiple regression model was fitted to data for 56 districts. Yield and the proportion of marketable surplus to output are the most important factors determining market arrivals in most regions. The effect of marketing cost and price of wheat do not appear to be important in most regions. |

| Sl.No. | Source | Year | Title | Author's | Key contents |
|---|---|---|---|---|---|
| 6 | *Agricultural Marketing.* 34(4): 26-29 | 1992 | Farm output and marketing behaviour of rural poor: some observations. | Pothuluru, C; Yadagiri, P | Found that, the ratio of marketed quantity to total household production was 35% for rice, 63% for pulses and 100% for groundnuts, while all sorghum production was retained for family consumption. A considerable proportion of the produce marketed was sold to money lenders/ traders immediately after harvest, at lower than market prices, in order to pay off debts. Cooperative marketing is needed to free small and marginal farmers from the exploitation and malpractices of middlemen. |
| 7 | *Agricultural development, price policy and marketed surplus in India study of Green Revolution region.* 128 | 1991 | Agricultural development, price policy and marketed surplus in India - study of Green Revolution region. | Chand, R | Found that, systems of output supply and factor demand equations were developed for major crops in the Indian Punjab, a Green Revolution region of India, and were estimated simultaneously within the framework of a profit function approach. Chapter 2 provides a brief review of related studies. Chapter 3 deals with the description of the data sample, variables and analytical model used in the study. |

# 3 CHAPTER

# THEORETICAL ORIENTATION

This chapter broadly deals with described framework used in this study in order to provide a theoretical base for the empirical investigation and guidance for the selection of relevant predictor variables as well as to envisage a set of hypothesis for testing. This chapter includes,

3.1 Descriptive study of pulse crop

3.2 The conceptual framework of the study.

## 3.1 DESCRIPTIVE STUDY OF PULSE CROP

### Pulses: The crop of ecology and economic value

Pulses are part of the legume family, but the term "pulse" refers only to the dried seed. Dried peas, edible beans, lentils and chickpeas are the most common varieties of pulses. Pulses are very high in protein and fibre, and are low in fat. Like their cousins in the legume family, pulses are nitrogen-fixing crops that improve the environmental sustainability of annual cropping systems. Pulses come in a variety of shapes, sizes and colours and can be consumed in many forms including whole or split, ground in to flours or separated into fractions such as protein, fibre and starch.

Pulses are a source of supplementary protein to daily diets based on cereals and starchy food for a predominantly vegetarian population and for those who cannot afford expensive animal protein. Pulses are a great tasting addition to any diet. They are rich in fibre and protein, and have high levels of minerals such as iron, zinc, and phosphorous as well as folate and other B-vitamins. In addition to their nutritional profile and links to improved health, pulses are unique foods in their ability to reduce the environmental footprint of our grocery carts. Put it all together and these sensational seeds are a powerful food ingredient that can be used to deliver the results of healthy people and a healthy planet. They also provide energy, essential minerals, vitamins and several compounds considered beneficial for good health. Their cultivation enriches soil by adding nitrogen, and improves the physical, chemical and biological soil properties.

Pulses are grown since ages in different parts of the world. They are well suited to diverse environments and fit in various cropping systems owing to their wide adaptability, low input requirements, fast growth, nitrogen fixing and weed smothering ability. Their short growing period and photo-period sensitivity make them suitable for crop intensification and diversification.

The demand of pulses is fast increasing, both in developed and developing countries, where they meet the minimum protein requirements of an increasing population turning to a vegetarian diet. Their productivity can be doubled by improved cultivars and by modern production technologies.

## Pulses in Historical agriculture:

Since most primitive mode of human life agricultural systems were evolved to fulfill the changing needs of human being. When the primitive man faced the problems in gathering his food in the form of edible fruits, seeds, roots leaves etc. and failed to have year-round supply, he entered the phase of hunting. Later he turned to animal husbandry i.e. protecting the herds from natural predators, guided to good grazing and care of water and so on. Further human interest grew in animal breeding, feeding and cropping of the herd or flock and protection from predators as well as diseases to the best of his ability. This made them to migrate periodically to sustain the feeding requirements of the herds and became nomadic which later settled around perennial water supplies. Modern agriculture, thus is originated for convenience in food gathering habit of human being which has been originated by propagating the useful plants near his own homestead from the seeds and vegetative propagation materials, collected from the experiences of pre-agricultural community. An organized form of this practice is nothing but Agriculture.

Human dietary considerations in the course of these developments had major impact on the selection during the domestication of pulses. Although the analytical profile of amino-acids of cereals and pulses was not available, the well-nourished community with balanced diet must have experienced the health benefits which led to use of complementary mixtures of cereals and pulses. There can be additional considerations of benefits to the domesticated animal herds from the use of pulses in feed and benefits in terms of soil health for perpetuating pulses in the agricultural systems, more particularly in arid and semi-arid tropical regions.

The word legume comes from *legere* meaning "to gather" which indicates that the seed were collected by hand instead of being threshed from the plants like cereals. Thus, legume has necessarily the characteristics of collecting seeds before it is shattered due to dehiscence of pods, which is very common in wild forms.

## Domestication of pulses:

Eevolution of pulses and concluded that selection against explosive dehiscence of pods which results in shattering of grains was the most important trait against which the selection was operated during the course of domestication of legumes for food. The species thus brought under cultivation has at least delayed dehiscence of the pod if not suppressed totally.

It has been well accepted that in evolutionary history of plants the inbreeders have been originated from out-breeders. The change from an allogamus to autogamous breeding system may sometime depend on simple genetically governed mechanisms like cleistogamy in Pigeonpea (Mahta and Dave, 1931; Datta and Deb, 1970; Saxena *et al.* 1987).With the exceptions of *Vicia faba* and *Phaseolus coccineus*, most of the pulses had already abandoned exogamy before domestication, although the floral structure is well adapted to cross pollination.

Similar to most of the domesticated agricultural plant species the morphological changes in terms of improved size of fruits and seeds are equally concerned to pulses. The increased leaf size without much rise in plant dry matter means plant less freely branched with fewer larger leaves which is likely to be occurred during evolution of such species through the change from many thin wiry stems/branches bearing numerous small leaves. Cowpea, chickpea and soybean are cited as another few examples in which a restriction of branching during evolution is accompanied by compensating increase in those parts (leaves, pods, seeds) and reduction in their number, finally resulting in a very compact, more or less erect growth form suited to better adaptation in the agricultural system.

Physiological changes associated with the developments have also gone through similar selection, may not be deliberate one. Moreover all these changes in domestication are related to ability of plant to yield the economic products, many times it may be seeds of desirable quality. Selection toward annual growth habit must have an important role in adoption of most of legumes, wild forms of which are seen to be perennial.

Under domestication of pulses there is every possibility to have selection pressure to adopt them to the changing environment. Kabuli chickpea recently placed in southern peninsular India is a unique example of selection of early flowering, fast growing in early seedling stage to suit to a short-mild winter environment.

## The nutritional value of pulses:

Pulses are basically dried edible seeds obtained from plants. A good source of proteins, they are a staple food in many parts of the world. In the tropical areas especially, pulses are one of the most important food groups after cereals. In fact, pulses are a good source of nutrition as they have significant quantities of both proteins and calories.

Pulses are low fat source of protein, with a high fibre content and low glycaemic index:

- Pulses are very high in fibre, containing both soluble and insoluble fibres. Soluble fibre helps to decrease blood cholesterol levels and control blood sugar levels, and insoluble fibre helps with digestion and regularity.

- Pulses provide important amounts of vitamins and mineral. Some of the key minerals in pulses include iron, potassium, magnesium and zinc. Pulses are also particularly abundant in B vitamins including folate, thiamine and niacin.
- Pulses typically contain about twice the amount of protein found in whole grain cereals like wheat, oats, barley and rice
- Finally, pulses are an important source of proteins and in most developing countries constitute the main source of protein for most populations.

In addition to contributing to a healthy, balanced diet, pulses nutritional qualities make them particularly helpful in the fight against some non-communicable diseases.

The World Health Organisation estimates that up to 80% of heart disease, stroke, and type 2 diabetes and over a third of cancers could be prevented by eliminating risk factors, such as unhealthy diets and promoting better eating habits, of which pulses are an essential component. Up to 80% of heart disease, stroke, and type 2 diabetes and over a third of cancers could be prevented by eliminating shared risk factors, mainly tobacco use, unhealthy diet, physical inactivity and the harmful use of alcohol. Pulses can help lower blood cholesterol and attenuate blood glucose, which are a key factors in the fight against diabetes and cardiovascular disease. Eating pulses as a replacement to some animal protein also helps limit the intake of saturated fats and increases the intake of fibres. Pulses have also been shown to be helpful in the prevention of certain cancers, because of their fibre content but also because of their mineral and amino-acid contents, in particular folate.

Pulses are included in all 'food baskets' and dietary guidelines. The World Food Programme (WFP) for instance includes 60 grams of pulses in its typical food basket, alongside cereals, oils and sugar and salt. Encouraging awareness of the nutritional value of pulses can help consumers adopt healthier diets. In developing countries, where the trend in dietary choices tends to go towards more animal based protein and cereals, retaining pulses is an important way to ensure diets remain balanced and to avoid the increase in non-communicable disease often associated with diet transitions and rising incomes.

Several studies have shown that legumes are been associated with long-lived food cultures such as the Japanese (soy, tofu, natto, miso), the Swedes (brown beans, peas), and the Mediterranean people (lentils, chickpeas, white beans) and that they could be an important dietary factor in improving longevity.

**Table 3.1. Nutrition values of different pulse**

| Nutritive Values of different pulses commonly used in India | | | | | | | | |
|---|---|---|---|---|---|---|---|---|
| Sl. No. | Pulses (100g) | Protein (g) | Fat (g) | Carbohdrate (g) | Energy (kCal) | Cruide fibre (g) | Calcium (mg) | Iron (mg) |
| 1 | Bengal gram | 17.1 | 5.3 | 60.9 | 360 | 3.9 | 202 | 4.6 |
| 2 | Black gram | 24 | 14 | 59.6 | 347 | 0.9 | 154 | 3.8 |

| Nutritive Values of different pulses commonly used in India | | | | | | | | |
|---|---|---|---|---|---|---|---|---|
| Sl. No. | Pulses (100g) | Protein (g) | Fat (g) | Carbohdrate (g) | Energy (kCal) | Cruide fibre (g) | Calcium (mg) | Iron (mg) |
| 3 | Cow pea | 24.1 | 10 | 54.5 | 323 | 3.8 | 77 | 8.6 |
| 4 | Field bean | 249 | 0.8 | 60.1 | 347 | 14 | 60 | 2.7 |
| 5 | Green gram (Whole) | 24 | 1.3 | 56.7 | 334 | 4.1 | 124 | 4.4 |
| 6 | Green gram (split) | 24.5 | 1.2 | 59.9 | 348 | 0.8 | 75 | 3.9 |
| 7 | Horse gram | 22 | 0.5 | 57.2 | 321 | 53 | 287 | 6.77 |
| 8 | Kesri dhal | 22 | 0.6 | 56.6 | 345 | 2.3 | 90 | 6.3 |
| 9 | Peas (green) | 7.2 | 0.1 | 15.9 | 93 | 4.0 | 20 | 1.5 |
| 10 | Peas (dry) | 19.7 | 1.1 | 56.5 | 315 | 4.5 | 75 | 7.05 |
| 11 | Rajmah | 22.9 | 1.3 | 60.6 | 346 | 4.8 | 260 | 5.1 |
| 12 | Red gram | 22.3 | 1.7 | 57.6 | 335 | 1.5 | 73 | 2.7 |
| 13 | Soyabean | 43.2 | 19.5 | 20.9 | 432 | 3.7 | 240 | 10.4 |
| 14 | Groundnut | 25.3 | 40.1 | 26.1 | 567 | 3.1 | 90 | 2.5 |

Source: www.fao.org

## Pulse's share to total food grain basket

Per cent share of pulses to total food-grain basket in the country in terms of area and production was 19.62 and 16.55 per cent respectively during 1950-51. This trend continued till 1960-61 and started declaration from 1970-71(after green revolution) due to no breakthrough in production technology of pulses in comparison to other commodities of food grains. At present, except the area stabilization, the production during 2012-13 has gone down to 7.13 per cent due to stagnation in productivity of pulses as compared to other commodities of food grains. Deceleration of Per cent contribution of pulses to total food grains has prompted the Ministry of Agriculture to vigorously pursue the NFSM-Pulses during the Twelfth Plan (2012-13 to 2016-17), a Centrally Sponsored scheme, in addition to ongoing ISOPOM scheme for all 14 pulse potential states. Food grains cover almost 62% of total gross cropped area comprising cereals 51% and pulses in about 11% in India. Further, among total pulses, the area under gram is 4%, arhar 2% and the other pulses in about 5% of gross cropped area. The net irrigated area in the country is 47% while the remaining falls under rain fed ecology. The pulses under irrigation are cultivated in about 37% of the area while 63% of pulses are grown under rain fed conditions.

**Table 3.2 Area Production and Yield of Total Pulses in India**

| Year | Area (Lakh hectares) | Production (lakhtones) | Yield (kg./hectares) |
|---|---|---|---|
| **2000-01** | 203.48 | 110.75 | 544 |
| **2010-11** | 264.02 | 182.41 | 691 |

| Year | Area (Lakh hectares) | Production (lakhtones) | Yield (kg./hectares) |
|---|---|---|---|
| **2011-12** | 244.62 | 170.89 | 699 |
| **2012-13** | 232.57 | 183.42 | 789 |
| **2013-14** | 252.18 | 192.55 | 764 |
| **2014-15** | 235.53 | 171.52 | 728 |
| **2015-16** | 249.11 | 163.48 | 656 |
| **2016-17** | 292.77 | 224.01 | 765 |

**Source: DES, 2016-17*- IIIrd Advance Estimate**

## Per capita availability of pulses in India

The per capita per day availability of pulses in 1951 was 60 g that dwindled down to a provisional level of 47.2 g in the year 2014. The per capita per year availability shows the same decreasing trend from 22.1 kg in 1951 to 17.2 kg in 2014. However, the increase trend shows during 2017 (P) both in per capita per day (52.9 g) and per capita per year (19.3 kg) respectively. This amply proves that increase in population growth affects the pulses availability on per capita basis (Table 3.3).

**Table 3.3. Per capita availability of pulses in India**

| Year | kg per capita per year |
|---|---|
| **2010** | 12.9 |
| **2011** | 15.7 |
| **2012** | 15.2 |
| **2013** | 15.8 |
| **2014** | 16.9 |
| **2015** | 16.0 |
| **2016** | 15.7 |
| **2017** | 19.3 |

**Source: Agricultural statistics at a glance-2014.**

**Table 3.4. Demand and supply status – production and import/export**

| Year | Production (lakh tonnes) | Import (lakh tonnes) | Export (lakh tonnes) | Total availability (lakh tonnes) |
|---|---|---|---|---|
| **2009-10** | 146.62 | 35.10 | 1.00 | 182.71 |
| **2010-11** | 182.41 | 26.99 | 2.08 | 211.48 |
| **2011-12** | 170.89 | 33.65 | 1.74 | 206.28 |
| **2012-13** | 183.43 | 38.39 | 2.02 | 223.84 |
| **2013-14** | 192.53 | 30.49 | 3.43 | 219.60 |
| **2014-15** | 171.52 | 45.85 | 2.22 | 215.20 |
| **2015-16** | 163.48 | 57.98 | 2.56 | 218.90 |
| **2016-17** | 224.01 | 66.09 | 1.37 | 288.70 |

Source: DGCI&S, Ministry of Commerce, Kolkata.

Table 3.5. Trend in Minimum Support Prices for Pulses (Rs/q)

| Crop | 2014-15 | 2015-16 | 2016-17 | 2017-18 | 2018-19 |
|---|---|---|---|---|---|
| Moong | 4600 | 4850 | 5225 | 5575 | 6975 |
| Urad | 4350 | 4625 | 5000 | 5400 | 5600 |
| Arhar | 4350 | 4625 | 5050 | 5450 | 5675 |
| Gram | 3175 | 3500 | 4000 | 4400 | 4620 |
| Musur | 3075 | 3400 | 3950 | 4250 | 4475 |

**Source: DGCI&S, Ministry of Commerce, Kolkata**

## Promotion of Pulse Production in India

The increase in population, decrease in per-capita land availability and stagnation in pulses production has created a gap in demand and supply of pulses. The government has made various efforts to bridge this gap like launching of Technology Mission in 1986 in order to reduce import and achieve self-sufficiency in production of particular pulse crop covered under the Mission. Pulses were brought within the purview of the Mission with the introduction of National Pulses Development Project in 1990-91. The centrally sponsored National Pulses Development Project was implementation in 30 states/UTs in the country up to the year 2003-2004 covering 356 districts. Implementation of the project helped in increasing the production of pulses from 12.86 million tonnes in 1989-90 to 15.23 million tonnes during 2003-04 with a marginal decline during 1999-2000 to 2002-03 due to occurrence of drought. Various schemes under the technology mission including National Pulses Development Project were merged into one Centrally Sponsored Integrated Scheme of Oilseeds, Pulses, Oil palm and Maize (ISOPOM) during 2004 for providing flexibility to the states in implementation of these schemes. The Government has introduced the National Food Security Mission in 2007-08, with the mandate among others, to increase production of pulses through area expansion and productivity enhancement in sustainable manner in the identified states. Accelerated Pulses Production Programme (A3P) is another step forward for vigorous implementation of the pulse development under the NFSM – Pulses. A3P has been conceptualized to take up the active propagation of key technologies such as Integrated Nutrient Management (INM) and Integrated Pest Management (IPM) in a manner that creates catalysing impact by assuring farmers of the higher returns from the identified pulse crop

## Pulses: The market and the network

Pulses are important component of diet in Indian, mainly for being one of the leading sources of protein for predominantly vegetarian society. The emphasis of the Government has, accordingly, been on increasing production of pulses through area expansion and adoption of improved technology so as to ensure the availability of pulses to the masses as per quantity recommended by ICMR. Equally important is the marketing of the pulses particularly in a situation created by the introduction of reforms in the sector and existing demand-supply gap in pulses. Shortage of supply of pulses has resulted in increase in prices, thereby pushing pulses out of the reach of poor household leading to a negative effect on their nutritional status (Reddy, 2004). The Indian food economy, since 1970, has also undergone major policy reforms, including trade liberalization leading to opening up of domestic pulse market to international trade. The market is very lucrative and of major significance to the world pulse economy (Agbola & Damoense, 2003) as has been evident by the increased imports of pulses in India in recent past. The marketing component is important to ensure remunerative prices to the farmers which will eventually work as an incentive for them to bring more area under pulses. The Government has provided price support to the framers through minimum support price of leading pulses. However, the marketing and price policy of the government should be the focus of various government initiatives mainly the extension programmes, so that the awareness may be created among farming community and farmer's may 14 respond more effectively to the various production incentives given by the Government (Savadatti and Narappanavar, 1998). Marketing can also help in inducing an element of incentive to farmer through participation in processing and distribution of pulses through direct marketing, farmers market or cooperative marketing to get higher share in the consumer's price. Marketing innovations like group marketing will help in improving the bargaining powers of small and marginal farmers.

## Constraints in pulse production

There are several constraints in pulse production. These crops are generally treated as second rate crops in comparison to paddy during kharif and wheat during rabi by the farmers. The first grade well fertile, irrigated lands are allocated for paddy and wheat and sub-standard un-irrigated land is generally left for these pulse crops. Farmers used to find good quality and best variety seed for paddy and wheat but generally grow their own saved seed for pulses. It indicates that there are less systematic efforts by the researchers to develop varieties of these pulses for different agro climatic zones as compared to wheat and paddy. As far as the fertilizers are concerned, farmers use almost all types of required fertilizers for wheat and paddy but hardly use any fertilizer for pulse crops. Farmers are unable to use recommended doses of fertilizers due to lack of soil testing laboratories in remote villages. The Government support may be required to establish at least one soil testing lab in the cluster of about ten villages. It will not only create employment opportunities but a lot of fertilizer may also be saved which is being used by farmers without knowing the actual requirement of the fertilizers. The study of ICRISAT reported that seed replacement rate in India is as low as 4% (Rao 2010).

The farmers generally sow pigeonpea and chickpea through broadcasting method despite informing them the advantages of line sowing. However, lack of machinery to maintain proper distance for sowing leaves no option to go for broadcasting method. It is suggested that ridge planting for proper plant population may be followed for higher yields. The agricultural engineers should develop machineries suiting to the requirement of pulses. Pod borer (Helicoverpa armigera) is very harmful insect of chickpea and reported to damage 20 to 40 per cent crop. The population of pod borer can be managed by intercropping of mustard and coriander or linseed and pheromone traps (Reena et.al. 2009). The losses in pigeonpea crop on farmer's field are about 20 to 100 per cent (Rao 1993). Crop damage by wild animals and high cost of input were the major constraint among other related constraints.

## Pulse Production and Sustainability

### Environmental benefits:

**Nitrogen fixation:** Nitrogen is the nutrient most commonly deficient in soils around the world, and is therefore the most commonly applied plant nutrient, often in the form of synthetic fertilizer. Legume crops, including pulses, have a unique role to play in the global nitrogen cycle, as they fix atmospheric nitrogen in soils. Pulses create a symbiotic association with rhizobia, a soil bacterium, enabling pulses to fix atmospheric nitrogen gas, which can make them self-sufficient in nitrogen, and enable them to grow in almost any soil without fertilizer inputs. Human impacts on the global nitrogen cycle from rapidly increasing fertilizer use and fossil fuel combustion starting in the 20th century have had strong negative effects, such as pollution into waterways and increased $N_2O$ emissions.

From 1960 to 2000, nitrogen fertilizer use increased by roughly 800%, with half of that being utilized for wheat, rice, and maize production (Canfield et al, 2010). Synthetic fertilizers provided close to half of all the nutrients received by crops globally during the mid-1990s, demonstrating both a large dependency on synthetic fertilizers, but also inefficient management of nitrogen in global agriculture (Smil, 2002). The IPCC estimates that nitrous oxide emissions contain roughly 300 times the global warming potential of carbon dioxide ($CO_2$), and application of fertilizer in agricultural production is a significant source of $N_2O$. Cereal crops such as wheat, rice and maize typically only utilize 40% of fertilizer applied, leading to significant waste and environmental impacts such as eutrophication of coastal waters and creation of hypoxic zones (Canfield et al, 2010). A survey of various field studies of nitrogen fertilizer uptake by rice, corn and wheat shows typical nitrogen efficiency to be less than 50%, with Asian rice averaging as little as 30%. The study also found nitrogen losses along the food chain to be significant, with synthetic fertilizers causing significantly more nitrogen loss due to volatilization, erosion and leaching into water (Smil, 2002).

However, introduction of pulses into crop rotations actively helps fix nitrogen in the soil, thus reducing the fertilizer requirements of the pulse crop itself, as well as the following grain crop. One long-term study (2001 – 2013) on the nitrogen fixation of thepulse crop itself found field pea, lupin or faba bean derived about 70% of nitrogen requirements from atmospheric nitrogen, while an average of 19 kg of nitrogen was fixed per tonne of pulse shoot dry matter. The study covered the geographic range of southern and central New South Wales, Mallee and Wimmera in Victoria, and the high-rainfall zone of south-eastern South Australia (Peoples et al, 2015). Systematic crop rotation based on incorporating pulses/legumes into maize-based systems to reduce synthetic fertilizer use, and optimizing the timing and amounts of fertilizer applied to crops are recognized as the two most important interventions to decrease nitrogen application (Canfield et al, 2010). Biological nitrogen fixation is a crucial alternative source of nitrogen, and can be enhanced along with other integrated nutrient management strategies such as animal manure and other bio solids, and recycling the nutrients contained in crop residues (Lal, 2004).

The higher available nitrogen to subsequent cereal crops is generally assumed to **benefityields of those cereal crops**, however this is not as well documented as the nitrogenfixation benefits. Findings in south-eastern Australia indicate strong evidence that the inclusion of legumes in cropping sequences results in higher available soil nitrogen for subsequent crops, with an additional 40 to 90 kg N/ha in the first year and 20 to 35 kg N/ha for the second year, as compared to continuous cereal sequences that do not include pulses (Peoples et al, 2015). Apart from the effects of additional fixed N that legume crops bring into systems, there are almost always beneficial yield effects from crop rotations with legumes. These positive effects on yield are probably related to disruption of the build-up of disease and pests that occurs when a particular crop is grown year after year, although this phenomenon is not yet well understood. Increases in grain yield in subsequent cereal crops have been documented in the Northern Great Plains of North America. Evidence in a Mediterranean environment demonstrated that vetch, faba bean and chickpea all resulted in significant yield surpluses and provided nitrogen credit to the subsequent unfertilized wheat crop, though vetch outperformed the other (better researched) pulses (Dalias, 2015). Experiences in Australia show increased yield and protein content in cereal and oilseed crops that are planted following pulse crops. The wide variations in the amount of nitrogen fixation that pulses can provide depends on the amount of biomass produced by the pulse crop (often varying with water, soil quality and non-N nutrient availability), whether the harvest removes a significant amount of the biomass, and the effectiveness of the legume-rhizobium symbiosis in fixing nitrogen. Planting a legume into soils already having moderate to high levels of soil nitrogen can also depress biological nitrogen fixation.

Nuances and differences in successive cereal crops must be noted. Findings in Australia comparing legume and fertilizer nitrogen indicate that recovery of legume nitrogen by a following cereal crop tends to be lower than top-dressed fertilizer(a difference of 20%between high and low ranges), but is not too dissimilar from fertilizer applied at sowing. This is due to slower release of mineral nitrogen from legume and pulse crop residues as they decompose. However, losses of nitrogen from the system are found to be usually lower from legume sources than from fertilizer, indicating the contribution of legumesto the maintenance of the long-term organic fertility of soils (Peoples et al, 2015).

The successive planting of different crops on the same plot of land, through crop rotations, helps soil fertility, the transfer of nutrients from one crop to the next, and helps to control weeds, pests and diseases. **Crop rotations have been practiced by farmers for thousands of years, yet maximizing the environmental, social and economic benefits of crop rotations is an ever-evolving science that must address many factors at the cropping system level**. The replenishment of nitrogen through the use of green manure, in sequence with cereals, is a common form of crop rotation. Including pulses in cropping systems has high relevance for improving the overall use efficiency of available nitrogen at the farm system level rather than at just the crop level. Rather than emphasizing individual elements of a cropping system, a focus on overall growing conditions, the crop mix, and the sequence of crop rotations is central to achieving both sustainability and productivity objectives, such as increasing nitrogen fixation through improved rhizobia-host plant symbiosis (van Kessel and Hartley, 2000).

- **Productivity improvements over area expansion:** An important goal in the sustainable use of land worldwide is to increase productivity on available croplands, while restricting agricultural expansion, which often occurs at the expense of forests and wetlands. Nine billion people will inhabit the planet by 2050. To avoid crop expansion and just meet projected 2050 crop needs by increasing production, it is predicted that crop yields would need to increase by an estimated 32% more from 2006 to 2050 than they did from 1962 to 2006 during the height of the 'green revolution (Searchinger et al, 2013).' However, reaching such increases in yields is highly unlikely. Pulses have a significant role to play in 'sustainable intensification,' yet, in developing countries, production increases have come primarily from expansion of cropping areas. The yield growth of pulses between 1980 and 2004 in developed countries was 2% per annum, while in developing countries, it languished at about 0.4% per annum (Nedumaran et al, 2015)

This large yield gap between developing countries and developed countries is of concern, and cannot be addressed by improved pulse crop genetics alone, but rather requires a range of interventions, some of which are further explored in the Africa case

- **Reduced greenhouse gas emissions:** The role of pulse crops in mitigating greenhouse gas emissions in agriculture production can be significant, and this is explored further in the Saskatchewan case study below. The primary reason for the benefits from pulses in lowering GHG emissions is due to lower fertilizer requirements, particularly given the large amount of energy used in fertilizer production. Up to 70% of the non-renewable energy used in Western Canadian cropping systems is due to the use of fertilizers, particularly nitrogen, and the inclusion of pulses in cropping systems reduces the need for fertilizer inputs. Pulses supply their own nitrogen and contribute nitrogen to succeeding crops (Lemke et al, 2007)

- **Pulses and livestock feed diversification:** Integration of legumes into livestock production systems has been shown to deliver multiple benefits, such as increased nitrogen supply while also increasing meat production. Globally, meat demand is expected to increase by 200 million tonnes per annum by 2050, with corresponding demand for livestock feed (Alexandratos and Bruinsma, 2013). In the northern Great Plains of the US and Canada, field pea has been promoted as a means to boost protein and energy in cattle feed. Field pea grain has been found to be highly digestible to cattle, but the starch fermentation and ruminal protein degradation rates are slower than for other common feeds. Field pea has been shown to increase dry matter intake by cows when included in the livestock feed ration, and also produces benefits when used as a binding agent for pelleting formula feeds.

## Social Benefits

- **Health and Nutrition:** Without a concerted effort to boost the production of pulses in developing countries, consumption of pulses may stagnate or decline, due to changing consumer preferences, failure to promote production of pulses, and a greater focus on increasing production and self-sufficiency in cereals (Alexandratos and Bruinsma, 2012). Historically, when observed declines in protein-rich pulses were not accompanied by increases in the consumption of livestock products, the result has been deterioration in the overall quality of diets, even if per capita dietary energy increased.

Pulses have declined in consumption levels globally and in particular among developing countries, perhaps best illustrated by China's significant decrease in consumption, from 30 g per capita per day in 1963, to only 3 g per capita by 2003 (Kearney, 2010). Over the last decade, developing countries, particularly in large Asian economies, have seen steady population growth, rising per capita incomes and continuous urbanization, which has been accompanied by increased protein intake relative to the traditional starches. The OECD/FAO Agricultural Outlook to 2024 indicates stagnant growth in food consumption in developed countries, but significant increases in developing countries, reflecting this increase in protein intake in developing countries. At the global level, total caloric intake is expected to rise, but with regional differences. Cereals will remain the most consumed agricultural product, with consumption expected to increase by almost 390 Mt by 2024, most of which is coarse grains. The largest expected growth in coarse gain consumption will come from demand for feed, accounting for 70% of the growth in consumption. Global meat consumption is expected to grow 1.4% per year, resulting in additional consumption of 51 Mt of meat by 2024 (OECD-FAO, 2015).

India provides a counterpoint to China, as pulses there provide an increasing source of protein, now accounting for almost 13% of overall protein intake (OECD-FAO, 2014). Pulses are an important contribution to a vegetarian diet, given their protein content. India is the largest pulse producer and consumer, and the country grows the largest varieties of pulses in the world, accounting for about 32% of the area and 26% of world production. Indian pulse crops include chickpea, pigeonpea, urd bean, mung bean, lentil and field pea, and production reached a record level of 18.4 Mt in 2012-13, up from 15 Mt in 2007- 08. Pulse crop yields have increased from 0.63 t/ha in 2007-08 to 0.79 t/ha in 2012-13, and annual yield growth is expected to outpace growth in production area, indicating better production efficiency. However, the average productivity of pulses in India still remains below the global average. It is expected that Indian production will not keep pace with demand and imports are anticipated to grow to 5.1 Mt by 2023 (OECD-FAO, 2014).

Fairly recent changes in the global human diet favouring more energy-dense foods rich in total and saturated fats are increasing rates of obesity, diet-related diseases such as diabetes, coronary heart disease and cancer. Public health and nutrition efforts seeking to promote healthier and more sustainable food production and consumption, must ensure a sufficient supply of staples and of micronutrient-rich foods without encouraging excessive consumption of energy-dense, nutrient-poor foods. Pulses and legumes are an important contributor of micronutrient-rich intake, along with fruits and vegetables (Kearney, 2010). Eighty-four percent of the protein in common bean is readily absorbed after consumption, and 94% of the protein from cowpea is available.

**Pulses have a role to play in combating cardiovascular disease and increasing healthy nutrition**. Worldwide, cardiovascular diseases are now the leading cause of death, and in the United States, is attributed to 1/3 of all deaths. While there is increasing awareness of this risk, and the role of balanced diets to decrease the risk, less than 1/3 of Americans consumes the 3 cups of legumes recommended per week by the Dietary Guidelines for Americans (Bazzano et al, 2011). A meta-analysis of ten randomized controlled trials from 5 countries sought to quantify the impact that consumption of non-soy legumes (navy, pinto, kidney, garbanzo and lima beans and peas such as split green peas or lentils) has on total cholesterol, high-density lipoprotein (HDL), low-density lipoprotein (LDL), very low-density lipoprotein (VLDL), and triglycerides. Findings indicate the **non-soy legume diet had a significant beneficial effect on serum cholesterol levels, thus reducing cardiovascular disease risk**. Both total and LDL cholesterol decreased, while HDL cholesterol did not change significantly, when non-soy legumes were supplemented. Further, findings indicate that the non-soy legume diet allowed for higher intakes of dietary total and soluble fiber (which lowers risk of coronary heart disease) and contained phytosterols, a component of plant cell membranes, which reduces blood cholesterol levels (*ibid*). The anti- inflammatory effects of beans and their contribution to the intestinal microbiome in the gut is increasingly being understood, particularly in among children in Malawi where findings indicate cowpea and common bean can reduce environmental enteric dysfunction (Manary, 2015).

- **Nitrogen and protein in global diets:** More dietary protein will be needed to eliminate disparities in diets between developed and developing economies. However, the nitrogen budget in global food and feed demonstrates **the**

**importance of nitrogen-rich and protein-rich plant foods**. About 70% of nitrogen in harvested food crops become available (after processing and losses) for human consumption, whereas meat and dairy production use large amounts of nitrogen. Nearly 7 kg of feed nitrogen is needed to produce 1 kg of edible nitrogen in meat, eggs, and dairy products. Thus, finding solutions to more efficient production of animal foods and damping the projected upward trend line of animal based foods is crucial to supply adequate nutrition to the world's growing population without any massive increases of nitrogen inputs (Smil, 2002).

In Canada, the US, and Europe, researchers, ingredient companies and food manufacturers are investigating **increased production and use of pulse protein fractions in manufactured food products,** in order to boost the nutritional quality of foods, and dietary fibre and starch, which can be used for food fortification and texture enhancement. In these processes, parts are derived from the whole seed, such as a split seed, hull or fibre, down to isolated starch and protein factions. These products are promoted as plant-based, sustainable, non-genetically modified and gluten-free. Companies such as Ingredion, AGT Foods, Burcon, Cosucra, and Nutri-Pea and others are creating fraction products, and finding ways to incorporate them into manufactured food products. Researchers note that wet fractionation uses large amounts of water and energy, and the functionality of the protein is compromised during processing. Dry fractionation is found to be more water and energy efficient, and retains functionality of the pulse protein.

Efforts to manufacture meat-like protein products is improving, with a recent soy-based meat analogue product being produced, with a thickness of 30 mm (Krintiras et al, 2016) . Efforts are underway to further identify how pea protein and other vegetable fibers can be used for this purpose.

- **Balancing livestock production and food security through pulses:** Livestock production benefits without food security conflicts are demonstrated via the integration of herbaceous legumes into the maize and upland rice systems in West Timor, Indonesia. After six years of participatory research, findings indicate significant benefits for local farmers. Farmers usually rely on low-quality native forages, sometimes supplemented, to feed their livestock. High rainfall levels in the monsoon season and climate variability provides risks that often result in feed demands outstripping supply and low rates of production, with up to 60% of the weight gained by livestock during the wet season lost during the late dry and early wet seasons. Herbaceous legumes were added either in annual rotation with a cereal or after wet- season cereal production has been completed, when land is traditionally left fallow. As cowpea, peanut, and mung bean are already occasionally intercropped with maize in the wet season, researchers chose to promote herbaceous legumes as a green manure or in rotation, so legumes would not further compete with the cereal crops for nutrients and thereby produce less forage. Research outcomes indicate that a) it is possible to add an additional crop into a traditional farming system without affecting existing food security, b) that feeding forage legumes to cattle during the late dry or early wet season resulted in increases in livestock weight, and c) that nitrogen provided by legumes improved maize production in subsequent crops. Critical to the success of integrating herbaceous legumes into an annual crop cycle was the recognition that water remaining in the soil as the main wet-season cereal crop matures is a resource that can be available for the subsequent dry-season production of herbaceous legumes, which is a period in which food crops are not traditionally grown (Nulik et al, 2013).

## Gender

Gender aspects of pulse production are particularly important in contexts where women can be involved in various stages of production and throughout the value chain.

## Economic benefits

Farmers in grain and oilseed production have found economic benefits from lower input costs and increased profits by including a pulse crop in their rotation. These benefits accrue mainly through enhancing the efficiency of nitrogen fertilizer use, reducing tillage and, in some cases reducing pesticide use. **Reduced and altered tillage practices reduce reliance on fossil fuels and lowers overall fuel bills**. No-till systems with pulses provide a basis for sustainable agricultural intensification, including integrated crop approaches. It is estimated that farmers save between 30-40% of time, labour and fossil fuels using no-till, compared to conventional tillage. Better nitrogen management requires less fertilizer inputs. There are variations, of course, in the economic returns experienced in different geographies and farm-types, and these are explored further in the case studies. Farmers likely also see the **long-term economic benefits (and avoided costs) of less soil, air and water degradation by adopting no-till practices and including legumes in their operations**.

India, which is the largest consumer of pulses globally. India sought to increase pulse production by 2 million tonnes by the end of the Eleventh Five Year Development Plan (2011-12), through implementation of the National Food Security Mission for Pulse Crops (NFSM). However, economic gains from pulse production could be far greater in India, and import tariffs, minimum price supports, and government support has largely under served the needs of India producing enough pulses to meet domestic demand and diversify farm incomes.

**Investments in pulse crop research is shown to have significant economic benefit**. TheCGIAR Research Program on Grain Legumes, a global alliance coordinating efforts across four CGIAR centres, estimated in 2012 that the net present value of gross benefits of its legume research and extension is estimated at US$ 4.5 billion, equivalent to US$ 535 million per year. Based on proposed activities to be undertaken by this CGIAR program, between 2014-2020, legume research was also projected to contribute to food security through increased availability of food (over 8 million tons), nutrition security from more availability of protein, and environmental benefits through biological nitrogen fixation (a fertilizer cost saving of US$ 418 million) . The CGIAR estimated that over 50% of the projected economic benefits of legume research and extension would accrue in South and South-East Asia and Sub-Saharan Africa, where most of the world's poorest communities are located (CGIAR, 2012).

**Adding pulses into livestock diets also has economic benefit, but appears dependent on market pricing and labour use efficiencies.** Findings in Western China indicate that livestock forage system intensification by incorporating a forage crop into grain-cropping systems increased average profits without increasing downside risks such as negative profit, crop failure, or livestock mortality. Forage vetch was the leguminous pulse crop, but forage oats and grain soybean were also incorporated. In contrast, replacing a grain crop with a forage crop in grain-cropping systems had a negative effect on profits, downside risk, and labour-use efficiency. Trade-offs between labour-use efficiency and profit were observed as forage intensification increased labour demands, however these effects were context specific (Komarek et al, 2014).

**The Rot in Indian Agriculture Policies and Economic challenges for Pulses in India**

Every aspect of Indian economic policies is influenced by all-pervading government control. Many times, they are also subject to massive leakages due to inefficiencies and corruption (for example, bogus ration cards, rotting grains and corrupt procurement in the public distribution system). The poor state of industrial development, inability to increase domestic defence production in spite of massive imports, diversion of government revenues from taxes on minerals, mismatch between demand for agricultural products and their production, and so on, are some examples. India has massive unemployment and underemployment. More are seeking fresh employment each year, consumption levels are low, and there are many poor people. Yet, growth in production of physical goods from agriculture and industry has, in five years since 2011, been modest. At constant prices, agriculture grew annually from 2011 in per cent at 1.5, 4.2, -9.2 and 1.1. Over these years, industry grew in per cent at 3.6, 5.0, 5.9, and 7.3; manufacturing at 6.0, 5.6, 5.5 and 9.5. There is heavy pressure of population on land. The average size of land holding was 1.41 hectares in 1995-96 and 1.15 in 2010-11.

India has more arable land than China. Indeed, it has the second-most arable land in the world. But it has very low productivity of crops per acre. When one takes paddy, as one example, and on comparing with China, India remains far behind. This is so for other crops as well and as compared to most Southeast Asian countries.

A good reason for the low productivity is the growing fragmentation of holdings and their decline in size. The high population pressure on small land holdings is on account of high rural poverty. Alleviation requires consolidation of land holdings by leasing, urbanization and the acquisition of rural lands for the purpose with adequate compensation. This will also reduce this pressure. Consumption habits are changing at all levels. Rice consumption per person per month in rural India was estimated at 5.98 kilogramme in 2011-12 compared to 6.38 kg in 2004-05 - a fall of 0.4 kg in seven years. In urban India, the fall in rice consumption between these two periods was 0.2 kg per person per month – from 4.71 kg to 4.49 kg per capita. A high proportion was taken at subsidized prices from the public distribution system. Per capita consumption of wheat in 2011-12 showed a slight rise since 2004-05 of about 0.1 kg per person per month in rural areas and a fall of 0.35 kg in urban areas. As with rice, the share of PDS purchase in wheat consumption has increased considerably, from 824 grammes to 901 gm in the urban sector. At the same time, and in contrast, 69 gm in the rural sector and 57 gm in the urban sector were contributed by split gram, whole gram, pea and *besan* bought at rising market prices. The four pulses - *arhar, moong, masur* and *urd* - also rose. So did consumption of vegetables, eggs and fruits. All these have seen rising prices and no price support to consumers.

Minimum support prices for cereals are increased almost every year and have been the same as procurement prices. But retail prices were not raised similarly. There is also government interference in the agricultural markets which cause the farmer to lose the margins made by middlemen. India has stimulated the production of cereals when the demand was dropping, while growing for other non-subsidized items. Rural households bought the cheap PDS grains and sold their own produce for government stocks, which are at unsustainably high levels without adequate storage. They might as well have been given free to the malnourished rural poor.

Productivity remained low because of falling land holdings. Improved seeds were not easily available. Genetic modification of food grains was prevented by environmentalists with no evidence of ill-effects. (Research has shown no ill-effects on humans; in some cases, there might be resistance to plant resistance.)

Fertilizers are subsidized, benefiting manufacturers and big farmers. Subsidies are relatively more for urea, leading to a mismatch in fertilizer use and inadequate productivity. Water availability has been shrinking as lakes and rivers become polluted and groundwater levels are used excessively.

Per-capita availability of fresh water has declined sharply, from 3,000 cubic metres to 1,123 cubic metres over the past 50 years *versus* the global average of 6,000 cubic metres. India needs to make judicious use of surface water and groundwater. Dams on rivers have robbed some of them of their usual water flow, while diverting the course of others. Urban effluents have also destroyed the portability of river water. Fifty-five per cent of India's total water supply is now groundwater. This has reduced levels across much of India. Growing water-intensive crops and using techniques like flooding for paddy have further depleted groundwater. Over 60 per cent of irrigation comes from groundwater. Nearly 30 per cent of urban water supply and 70 per cent of rural water supply come from groundwater. We need a rational water policy and less populism. Massive subsidies on equipment and electricity required to mine groundwater have accentuated its use to the financial detriment of the power sector. Free or heavily subsidized flat rate electricity tariffs helped bring the Green Revolution. They have now become the norm. The result is indiscriminate use of groundwater. Each of the issues mentioned is the result of government policies or their absence. Land legislation encourages fragmentation of holdings. There is little encouragement for leasing. Land acquisition for urban areas and for factories has been a corrupt business, and there is little incentive for many farmers to give up their land and move away. Government procurement policies and the public distribution system have stimulated rising production of rice and wheat. Little has been done to help production, storage and marketing of other items like vegetables, pulses, fruits and eggs (such as cold storages, improved varieties, and so on). There is little sign of a rational water policy: indeed, it is the opposite. Electricity tariffs encourage a rise in the use of groundwater. There is no policy on groundwater. Neither is there a limit on what water intensive crops can be grown with it. There is no concerted recharging of groundwater. State governments follow irrational water-pricing policies. Cleaning up river and lake water so that it is available to supplement water supply in case of drought is uncommon. Little is being done to spread better agricultural practices and scientific developments that benefit farmers.

Agricultural productivity can be increased even on small holdings, but there is little attempt, except by nongovernmental organizations, to disseminate these practices. In Israel, these have reduced the use of water.

There are vineyards producing good wine in the Negev desert. Groundwater use is not metered in India as it should be. Fertilizer subsidies do not benefit the small farmer. Rational policies, which ensure that subsidies reach the farmer and do not encourage unbalanced use of fertilizers, are urgently needed. Environmental agitations against genetically modified seeds should be countered by scientific evidence so that more productive seeds can be used. Agricultural policies have developed into a messy package. They need a thorough overhaul. Our political masters are unable to even think of it.

India is the largest consumer of pulses, but government subsidies and price controls in the agricultural sector created distortions that affected domestic production. Government subsidies for fertilizer and water promoted grains and oilseeds rather than the mix of support necessary to promote pulses. Protectionist policies in the 1970s and 1980s were reversed in the 1990s, when government reforms sought to remove import restrictions and lower tariffs on agricultural products. However, with the exception of Basmati rice and durum wheat, external trade in all major crops was regulated, and imports of most crops occurred only through government agencies. Pulses were treated differently, with import tariffs on pulses reduced gradually and abolished by 1996. The hope that domestic pulse market liberalization would increase imports did not materialize. Rather, the share of total pulse imports in total merchandise trade declined after market liberalization (Agbola, 2004).

Further, in India, minimum support prices have been established as one of the policy instruments used to improve the economic viability of farming, stabilize commodity prices, and enhance food security by diversification into oilseeds, pulses, livestock and fish. However, **minimum support prices for pulses have not demonstrated the same results as those for rice**. Prices for pulses were increased between 2008-09, at a rate higher than that for food grains, but that did not translate into larger areas planted under pulses, and this is attributed to the risks associated with pulse cultivation. In comparison, paddy cultivation does not carry such risks, and farmers are assured of procurement by government agencies, whereas this is not the case for pulses (OECD-FAO, 2014).

## 3.2 THE CONCEPTUAL FRAMEWORK OF THE STUDY

### Technology and socialization

Many a time we make mistake in understanding the differences between concept and commodity, need and devices to meet the need. a bag of fertilize thus presents some input, not the concept of nutrient management for getting desired yield a tractor, on the other way, is just a machine to harrow the land, not exactly in the concept to get the drudgery elimination substantially and done the work with less of error.

Then what is left is the socialization process that would combine the with commodity, techniques with tools, technology with culture. After having a technology socialized we can expect a complete social process in place. So, technology socialization process can be conceptualized as a comprehensive process through with a technology, the combined concept and commodity, has to undergo accommodation, assimilation and acculturation process and ultimately it would be transformed into a social character e.g a TV set in an electronic shop is just a commodity after being purchased and placed in a drawing room of any home, has become ingredient of family process. The family member now start socialization with this TV set by ascribing family status, taking it as marriage negotiation item, getting informed and entertained, deriving pride and galore by processing it and so on. This kind of commodity inflow into a social system through acculturation is called **creolization.**

Acculturation has been happening to every farm family with any exotic technology has been attempted to be diffused in a social system. Once a technology is being interjected, one may adopt, reject or reform it and in the passage of its subsequent adaption,one may discontinue or reinvent it. This whole lot of process undergoes,again, a social osmosis process.

- A social osmosis is basically a cultured screening process for desired assimilation and also a decided rejection over a proposed technology .it is just like a socialization process that a new born baby undergoes, some he/

she pick up and some he/she throw away but it is the entire process he/she undergoes an experiential learning process.

- In our cliché and conventions, we seldom take logic of reject and go happy to brand the rejecter a 'laggard' a category of non-adopter to be branded as an offender for not adopting anything prescribed by the experts.
- A technology socialization process logically includes all possible outcome or responses to technology prescription,i.e adoption, rejection, discontinuance, reinvention, elimination too. These all being done by farmer (or by any individual in this world) to get his existence adapted to the change process through a perfect thought process concluding in most intelligent manner a well. in most case we are not enough ready to catch up with this thought process happening in the mindset of a farmer in congruence with the change process in market, climate, socialechelons, groupdynamics,leadership, policies in all levels viz micro,meso and mega.

In understanding the technology socialization process, some of sociology terms need to be conceived interiliac.

**Socialization**: it is a process to provide an individual with skill and habits necessary for participating within their own society; a society itself is formed through a plurality of shared norms, customs, values, tradition, social and cultural continuity are attained.

There are different type of socialization:

**Primary socialization:** Primary socialization occurs when a child learns the attitude, values and actions appropriate to individuals as members of a particular culture.

**Secondary socialization:** Secondary socialization refers to the process of learning what is appropriate behavior as a member of a smaller group within the larger society.

**Development socialization:** Development socialization is the process of learning behavior in social institution or developing your social skills.

**Anticipatory socialization:** Anticipatory socialization refers to the process of socialization in which a person rehearses for future positions, occupation and social relationships.

**Re-socialization:** Re socialization refers to the process of discarding former behavior patterns and accepting new ones as part of a transition in one's life. This occurs throughout the human life cycle (Schalter and Lamm,1992)

Resocialization can be an intense experience, with the individual experience a sharp break with their past and needing to learning and be exposed to radically different norms and values.

**Gender socialization:** Gender socialization is the learning of culturally defined gender roles.

**Racial socialization:** Racial socialization refers to the process of learning one's culture and how to live within it bat refers more specifically to the socialization of ethnic minority group.

**Natural socialization:** Natural socialization occurs when infants and youngsters explore, play and discover the social world around them.

**Planned socialization:** Planned socialization occurs when other people take action designed to teach or train other –from infancy on natural socialization is easily seen when looking at the young of almost any mammalian species. planned socialization is mostly a human phenomenon, and all through history, people have been making plans for teaching or training others. Both natural and planned socialization can have good and bad features. it is wise to learn the best features of both natural and planned socialization and them into our lives.

**Positive socialization**: Positive socialization is the type of social learning that is based on pleasurable and exciting experience. We tend to like the people who fill our social learning process with positive motivation, loving care and rewarding opportunities.

**Negative socialization**: Negative socialization occurs when others use punishment, harsh criticisms or anger to try to 'teach us a lesson' and often we come to dislike both negative socialization and the people who impose it on us.

## Agent of socialization

Agent of socialization are the people and groups that influence our self-concept, emotion, attitude, and behavior.

1. The family: Family is responsible for the youth and, among other things, determiningone's attitude toward religion and establishing career goals.
2. Education: education is the agency responsible for socialization group of young people in particular skill and values in society.
3. Religion: religions play a major role in socialization in that context often synonymous with indoctrination
4. Peer group: peers refers to people who are roughly the same age and or who share other social characteristics (e.g student in college class)
5. The mass media
6. Other agents: work place, public institution.

| | |
|---|---|
| Primary socialization | Farmers start learning the benefits of a technology |
| Secondary socialization | Farmers start splitting the general benefit of the technology in term of his own farm situation. |
| Resocialization | Farmer reject a conventional technology to adopt a better alternative and get ready to absorb the shock of change. |

## Axiomatic structure for socialization process

The whole of the theorization process as well as approach has such been structure as to accommodate the following axioms:

| | |
|---|---|
| Axiom -1 | Post adoption behaviour could be delineated as the most important and crucial part of the discourse rather than the adoption process itself in providing logical support to describe **"socialization of technology"** |
| Axiom-2 | The elements as included in the echelon of the so called adoption process in the form of awareness ,interest, evaluation, trial ,adoption are not only skeptic or discrete but also get somewhere confusing or inappropriate while the entity the adoption behaviour in term of technology dissemination is in concern. |
| Axiom- 3 | An all comprehensive or resilient concept like **'socialization of technology'** is an alternative paradigm in itself to include and encompass all possible pre facto and post facto psycho social event to analysis the human behavior while it is being exposed or rendered interactive with a technology option or basket of techniques. |

The present study is logically based on the empirical elements of technology socialization process and all the post adoption phenomena like rejection, discontinuance, reinvention, etc., had comprised the crucial concept element towards estimating the technology socialization process as a whole, rather it's not a conventional way of studying technology gap, adoption, adoption level and the stuffs like.

CHAPTER 4

# RESEARCH METHODOLOGY

The deliberation on the methodology has been made to understand the concept, methods and techniques which were utilized to design the study, collection of information, analysis of the data and interpretation of the findings for revelation of truths and formulation of theories. This chapter deals with the methods and procedures which are used in the present study and presented in following eight sections:

4.1 Locale of Research

4.2 Sampling Design

4.3 Pilot Study

4.4 Variables and their measurement

4.5 Preparation of Interview Schedule

4.6 Pre-testing of Interview Schedule

4.7 Techniques of Data Collection

4.8 Statistical Tools used for Analysis of Data

## 4.1 LOCALE OF RESEARCH

### 4.1.1 Selection of District, Subdivision Block and villages:

The state of West Bengal consists of 19 districts; Nadia is one among this 19 districts which is again constituted by 18 blocks is selected purposively for the present investigation (Purposive sampling scheme). Two blocks from Nadia district are selected using purposive sampling scheme namely Chakdaha (consists 17 grampanchayats) and Haringhata (consists 10 grampanchayats). One grampanchayat from each selected blocks are then purposively selected, namely

Routari (under Chakdah block) and Fatepur (under Haringhata block). One village is selected from each grampanchayats chosen i.e. Maheswarpur (under Routari Grampanchayat) and Panchkahoniya (under Fatepur Grampanchayat) for the present study again using purposive sampling scheme. All the above said purposive sampling is done keeping in mind the following conveniences:

1. Ongoing International Fund for Agricultural Development project (IFAD Project) which covered pulse in the study area.
2. Under the supervision of university some farmers in the selected villages for the present investigation have already adopted pulse production technology. Hence there is ample scope for collecting relevant data needed for the study.
3. A number of progressive pulse growers are very much available in these villages and they have intensely adopted pulse cultivation.
4. The proximity of the study area to the investigator of the present study is very close and easily accessible in terms of transportation.
5. Availability of relevant and important information from the Assistant Director of Agriculture (A.D.A) and Krishi Projukti Shayak (K.P.S), etc.

## 4.2 SAMPLING DESIGN

Purposive as well as simple random sampling techniques were adopted for the present study. It may be termed as multistage random sampling procedure. The district, blocks and villages were purposively selected for the study. The next phase of study was to select the respondents. Farmers who were cultivating pulse in their own field or having tenancy status were selected as respondents. A list of such respondents from each village was collected from the village head and the village Gram Panchayat office. 150 farmers were finally selected randomly from the list that was collected from the village head and the village Gram Panchayat office. Here 75 farmers were selected randomly from each village selected. The respondents were interviewed through personal interview methods with the help of structured schedule which was developed for the study.

## 4.3 PILOT STUDY

Before taking up actual fieldwork a pilot study was conducted to understand the area, its people, institution, communication and extension system and the knowledge, perception and attitude of the people towards climate change concept. An outline of the socio-economic background of the farmers of the concerned villages, their opinion towards different types of technology socialization process, innovation-decision process, discontinuance, disagreement, conflict, rejection, dissonance, reinvention and confusion helped in the construction of reformative working tools.

The components of pilot study are:

- General information;
- Specific information;
- Prevalence of variables;
- Body languages of the prospective respondents;
- Access to physical location;
- The type, level and intensity of responsiveness;
- Related information including Agriculture, Animal husbandry, Fishery etc.

**Sampling Techniques and Sampling Design**

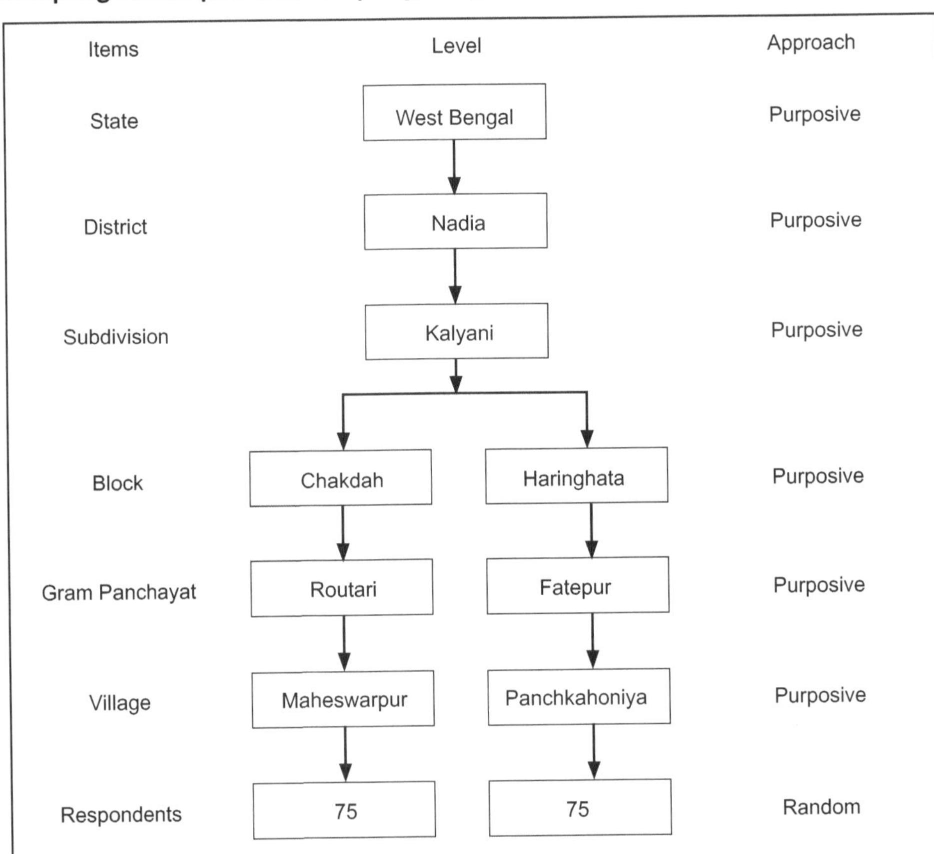


## 4.4 VARIABLES AND THEIR MEASUREMENTS

Several researchers pointed out that the behaviour of an individual was understood more in depth if one has the knowledge of some variables, which comprised the constructed world of reality within which an individual received the stimuli and acts. The socio personal, agro economic, socio-psychological and communication variables are such type of variables, which determine the behaviour of an individual. Appropriate operationalization and measurement of the variables help the researcher to land upon the accurate conclusion. Therefore, the selected variables for this study had been operationalised and measured in following manner.

Variables in the present study have been categorized into two main categories.

1. Independent variables.
2. Dependent variables.

## Independent Variables

The variables and empirical measurements.

**1. Age ($X_1$):** In all societies, age is one of the most important determinants of social status and social role of the individual. In the present study, the number of years rounded in the nearest whole number the respondent lived since birth at the time of interview, was taken as a measure of age of the respondent.

**2. Education ($X_2$):** Education may be operationalized as the amount of formal schooling attained/literacy acquired by the respondent at the time of interview. Education is instrumental in building personality structure and helps in changing one's behaviour in social life.

To quantify the educational status of the respondents, the scoring system followed by Pareek and Trivedi (1964) with modification in their socio economic status was used the scoring was as follows:

| Level of education | Score |
|---|---|
| Illiterate | 1 |
| Can read only | 2 |
| Can read and write | 3 |
| Primary | 4 |
| Middle school | 5 |
| High school | 6 |
| Higher secondary | 7 |
| Graduate | 8 |
| Post graduate | 9 |
| Other | 10 |

**3. Family Size ($X_3$):** Family size was operationalised as the member in the individual farmer's family. To quantify the family size, the score was given according to **Pareek and Trivedi (1964)** with modification like.

Total member in the family (actual member of family) : ................................

**4. Family education ($X_4$):** Family education status helps in providing the opportunity to the farmer to be in contact with outside world and hence seek new ideas and information. Family education status was operationalised as the average educational score of the family. The method followed by ray (1967) in computing the family education status was also followed in the present study. In this method the educational achievement of each number of the family was noted and scored, suggested Pareek and Trivedi (1964) with modification in the item 'education' of the socio economic status scale. The total score of a family on education was then divided by the 'effective family size' to get the education status of the family. The effective family size was obtained by subtracting the number of member up to 4 years of age from the total number of member in the family. The formula for computing family education status is

$$\text{FES} = \frac{\text{Total education score of a family}}{\text{Effective family size}}$$

5. **Area under pulse cultivation($X_5$):** Area under cultivation in this study refers to the total land available to the farmers where pulse is cultivated. The area under pulse cultivation was measured in bigha.

6. **Farm size and technology adoption ($X_6$):** Farm size is a measure of farm business operational farm size may be defined as a tract of land of possessed by an individual for the purpose of growing crop. different research worker had tried to measure farm size in different ways. in the present study, actual area under cultivation in bigha was taken as measure of farm size. This measure was adopted because of the obvious reason of accuracy under the following heading.

a. Homestead land + water body + orchard + crop land

b. Leased in land

c. Leased out land

7. **No crop diversity ($X_7$):** No of crop diversity has been operationalised as the proportion of no of crop grown to total farm size. The no of crop diversity was calculated by formula:

$$\text{No crop diversity} = \frac{\text{No of crop grown}}{\text{Total farm size}}$$

**8. Annual income (Rs/year/capita) ($X_8$):** The Annual Income of a person is an important parameter to assess the economic status of the person in the society. It was operationally defined as the gross income from all the viable source of income in a single year. It is measured in terms of rounded of rupees. The gross income is constituted by the total income generated from agriculture, dairy, poultry, fishery enterprises, business and services. In the present study it has been calculated with the formula as follows.

$$\text{Annual Income} = \frac{\text{Total income in a year}}{\text{Family size}}$$

**9. Risk Orientation ($X_9$):** In the present study, the risk orientation of farmers was measured with the help of risk preference scale developed by Supe (1969) with modification. Supe defined risk preference as the degree to which a farmer was oriented towards risk and uncertainty and had the courage to face the problems in farming. The scale consisted of six items

The items were rated in out of 10 (1 -10) point scales, farmer's preference was asked to give score out of 10(1-10). After obtaining the respondent's 'score out of 10(1-10) choices for each of the statements, the scoring was done by summing up score of all statement divided by total no. of statement (All together 6 statement). Following Hobbs and others (1964) this procedure was used by Supe (1969) with modification assuming that greater the total score of the subject on any one out of 10(1-10) point scale, the more positive his attitude or greater value is placed on that dimension.

10. **Scientific Orientation ($X_{10}$):** In the present study scientific orientation was operationalized as the characteristic of individual, which made him to trust and rely on ideas and practices developed through scientific research. This variable was measured with the help of scale, developed by Supe (1969) with modification. The scale consisted of six statements and each statement was of 10 (1-10) point scale and asked to the farmer to give preference score out of 10 (1-10). The score for each respondent in scientific orientation scale was obtained by summating the score for each statement divided by total statement (All together 6 statement).

11. **Management Orientation:** Samanta (1977) defined management orientation as the degree to which a farmer is oriented towards scientific farm management comprising of planning, production and marketing functions in his farm.

Management orientation is operationalized as the degree to which the individuals are oriented towards managing their income generating activities regarding planning, production and marketing function.

The management orientation scale developed by Samanta (1977) with modification was adopted for this study, consisted of six statements. The statements were rated in out of 10(1-10) point scale. The scoring was done by summing up score of all statement divided by total no. of statement (All together 6 statement).

**A. Planning ($X_{11}$)**

**B. Production Orientation ($X_{12}$)**

**C. Market Orientation ($X_{13}$)**

12. **Farmers attitude towards Pulse crop cultivation ($X_{14}$):** Thurstone (1946) defined attitude as the degree of positive or negative effect associated with psychological objects like symbol, phase, slogan, institution, ideas etc towards which people can differ in varying degrees.

The scale consisted of eleven item. The item was rated in out of 10 (1-10) point scale response categories and farmer's preferences were asked to give score out of 10 (1-10). The original scale was developed by Singh (1969) was used for measuring the attitude of farmers towards high yielding programme and had to be suitably modified for application in case of measuring the attitude of the farmers towards pulse cultivation. Care was taken to see that the modifications were minimum and the idea of each of the item remained unchanged. The modified scale was adopted for the present study after pre-testing. The score of the variable for an individual was obtained by summing up the scores of the eleven item divided by total no of item (all together 11 items)

**13. Knowledge level of farmer towards cultivation of pulse crop ($X_{15}$):** Knowledge is generally understood as an intimate acquaintance of an individual with facts English and English (1958) defined knowledge as a body of understood information possessed by an individual or by a culture. Knowledge was defined as the behaviour or test situations, which emphasize the remembering, either by reorganisation or recall of ideas material or phenomenon (Bloom *et al.* 1956).

The knowledge level was tested through a questionnaire of 15 questions, prepared on the information collected from the state department of agriculture, I.C.A.R and a report published by NABARD. Each correct answer was assigned the score of "1" and "0" was assigned for the wrong answer. The score was obtained by adding scores of all item.

**14. Knowledge about insecticides ($X_{16}$):** In this study knowledge was operationalised as those behaviours and test situations, which emphasized the remembering either by recognition or recall of idea, material, phenomenon. The variable indicated the extent of knowledge of the respondents about the agricultural practices regarding plant protection (insecticides). It was measured by using the scale of **Sagar** (1983) with modification after slight modification in case of item. Care was taken to see that the modification was minimal and the idea of each of the item remained unchanged.

The modified scale was adopted for the present study after pre testing. In this scale there were six items with dichotomized statements. The total score was obtained by adding scores of all items divided by total no. items.

**15. Knowledge about fungicide ($X_{17}$):** In this study knowledge was operationalised as those behaviours and test situations, which emphasized the remembering either by recognition or recall of idea, material, phenomenon. The variable indicated the extent of knowledge of the respondents about the agricultural practices regarding plant protection (fungicide). It was measured by using the scale of **Sagar** (1983) with modification after slight modification in case of item. Care was taken to see that the modification was minimal and the idea of each of the item remained unchanged.

The modified scale was adopted for the present study after pre testing. In this scale there were six items with dichotomized statements. The total score was obtained by adding scores of all items divided by total no. items.

**16. Knowledge about weed control ($X_{18}$):** In this study knowledge was operationalised as those behaviours and test situations, which emphasized the remembering either by recognition or recall of idea, material, phenomenon. The variable indicated the extent of knowledge of the respondents about the agricultural practices regarding plant protection (fungicide). It was measured by using the scale of **Sagar** (1983) with modification after slight modification in case of item. Care was taken to see that the modification was minimal and the idea of each of the item remained unchanged.

The modified scale was adopted for the present study after pre testing. In this scale there were two items with dichotomized statements. The total score was obtained by adding scores of all items divided by total no. items.

**17. Knowledge about IPM practice ($X_{19}$):** In the present study, the knowledge about IPM was operationalised as the working idea material of the respondent regarding integrated pest management. The variable indicated the extent of knowledge of the respondents about IPM. It was measured with the help of the scale developed by Choudhury (2000) with modification. In the present study sixteen item were dichotomized type question with 'yes' or 'no' type answer having score '2' and '1' respectively. The score of variable for an individual was obtained by summing up the score of the sixteen items divided by total number of item. (All together 16 items)

**18. Farmers attitude towards IPM programme ($X_{20}$):** Thurston (1946) defined attitude as the degree of positive or negative effect associated with psychological object like symbol, phrase, slogan, phrase slogan person, institution ideal or ideas etc toward which people can differ in varying degree.

The scale consisted of fourteen item. The items were rated in out of 10 (1-10) point scale response categories and farmer's preference were asked to give score out of 10 (1-10). The original scale developed by Singh (1969) with modification was used for measuring the attitude of farmers towards integrated pest management programme and had to be suitably modified (worded) for application in case of measuring the attitude of farmer toward IPM practices. Care was taken to see that the modification was minimal and idea of each of the item (statements) remained unchanged. The modified scale was adopted for the present study after pre testing. The score of the variable for an individual was obtained by summing up the score of the sixteen item divided by total number of item (all together 14 items).

**19. Attitude towards adoption ($X_{21}$):**

**20. Attitude towards discontinuous ($X_{22}$)**

**21. Attitude towards rejection ($X_{23}$)**

In the present study the attitude towards adoption had operationalised as the predisposition to behave in a consistent way in terms of accepting the given agriculture technologies.

Accordingly, the attitude towards discontinuance was defined as the predisposition to reject certain agriculture technology after its previous adoption.

Similarly, the attitude towards rejection was operational as the predisposition of an individual not to adopt certain agricultural technologies.

In the present study all these three variables were measured with the scales developed by Acharya and Pradhan (2005). The process of scale construction had been slightly modified for present study. The scales were included in the final schedule for data collection with modification of taking 10 (1-10) point scale and farmers preference were asked to give score out of 10 (1-10).

The score of the variable for an individual was obtained by summing up the scores of the items divided by total number of items.

**22. Productivity or yield (kg/Bigha) ($X_{24}$):** Total production of pulse crop per bigha

**23. Gross return (Rs/ Bigha) ($X_{25}$):** Gross income is total income from pulse of cultivation

**24. Utilization of source of Information ($X_{26}$):** Extension contact refers to the extent of interaction pattern (Localite and Cosmopolite) and exposure of an individual to different sources of information.

To measure the degree of utilisation of mass media sources, each respondent was asked to indicate on a 4point continuum divided by all item developed by **Singh (1993)** as how much information they got about improved agricultural practices from each of the sources. The sources were further classified into three parts, comprising of-

1. Mass Media
2. Personal Cosmopolite
3. Personal Localite

The scoring for the response categories were as follows.

| Items | Score |
|---|---|
| Most often | 4 |
| Often | 3 |
| Some times | 2 |
| Never | 1 |

**25. Training received ($X_{27}$):** In this study training received refers to whether an individual farmer has attended the farmers training programmes or not during last three years (from the date of interview) and if so for how many days. From the farmer training programme, farmers generally gather some new knowledge regarding production and infrastructure available. This measurement was previously done by De and Rao (2000).

It may be assumed that farmers training programme is one of the means by which the desired changes in knowledge and skill of farmers can be brought out. The individuals 'attitude and motivation, influenced by training programme will lead to their adoption behaviour.

In the present study, the attendance of farmers in the training meeting arranged for them, was taken as a measure of participation in farmers training programme. The number of days the respondent attended the training programme in the last 3 years was taken as a measured of their received.

## DEPENDENT VARIABLES.

**1. Socialization of pulse enterprise ($Y_1$):** In the Socialization of pulse enterprise, the score was obtained by summing up the score of Extent of adoption, Extent discontinuance, Extent of rejection divided by total no. of items (All together 12 items were considered the period of adoption, level of adoption, period of discontinuance, level of discontinuance, period of rejection and level of rejection).

a. **Extent of Adoption:** The adoption phenomenon was operationalised as a decision by an individual to get on with an innovation for protractive period in a given social system. In the present study, the extent of adoption was measured and the extent of adoption score was determined through a relevant modification and revision of 'adoption scale' developed by S. N. Chattopadhyay (1963) by considering the potentiality, extent, time and consistency. Here the researcher calculated the extent of adoption score by following the given formula:

$$A_E = \frac{\Sigma_{i=1}^{n} Y_{Ai} * L_{Ai}}{n_A}$$

Where $A_E$ = Extent of adoption, YA, = The period of adoption for •t items of adoption, for i[th] item of adoption, $L_{ai}$ = The level of adoption for i[th] item of adoption. In this study:

Level of adoption

$$= \frac{\text{Area covered by innovation}}{\text{Potential area that could have been covered by the innovation}}$$

$n_{A}$ = Total number of item of adoption.

b. **Extent of Discontinuance:** The extent of discontinuance was measured after operationalising the discontinuance phenomenon. The discontinuance was operationalised as a unique adoption decision where in an innovation was adopted in its first exposure by a potential adopted and dropped out due to some situational factors. According the extent of discontinuance was also measured through a relevant modification and relevant modification and revision of 'adoption scale' developed by **S.N Chattopodhyay (1963)**. As it previously mentioned that discontinuance was a adoption process the researcher had followed the given formula to calculate the extent of discontinuance score:

$$D_E = \frac{\Sigma_{i=1}^{n} Y_{Di} * L_{Di}}{n_D}$$

$D_E$ = Extent of discontinuance

$Y_{Di}$ = Priod of discontinuance for ith item of discontinuance.

$L_{Di}$ = level of discontinuance for ith item of discontinuance.

In this study.

level of discontinuance (Ld)

$$= \frac{\text{Area impact through discontinuance}}{\text{Apotential area that could be impact by innovation}}$$

c. **Extent of rejection**: Rejection is a more revealing and cognitive counter phenomenon of adoption .it was operationalized as the decision not to adopt an innovation or conventional one .as it is a counter process of adoption the extent of rejection was also measured through a relevant modification and revision of adoption scale develop by SN Chattopadhyay (1963).in the present study the extent of rejection score was calculate by following the given formula:

$$R_E = \frac{\Sigma_{i=1}^{n} Y_{Ri} * L_{Ri}}{n_R}$$

$R_E$ = Extent of rejection

$Y_{Ri}$ = Period of rejection for $i^{th}$ item of rejection

$L_Ri$ = Level of rejection for $i^{th}$ item of rejection.

In this study

$$\text{Level of rejection (Lr)} = \frac{\text{Area impact through rejection}}{\text{Apotential area that could be impact by innovation}}$$

**2. Socialization through replacement ($Y_2$):** It has been calculating in term of percentage by subtracting previous crop (bigha) from present crop (bigha) divided by total land and multiplied by 100.calculate by following formula:

$$Y_2 = \frac{\text{Present crop (bigha)} - \text{Previous crop(bigha)}}{\text{Total land}} \times 100$$

**3. Socialization through reinvention ($Y_3$):** It has been calculated in terms of percentage by following formula

$$Y_3 = \frac{\text{No of addition of technology} + \text{No of elimination of technology} + \text{No of modification of technology}}{3} \times 100$$

4. **Marketed Surplus($Y_4$):** Marketed surplus is that quantity of the produce which the producer farmer actually sells in the market. Irrespective of his requirement for family consumption, farm needs and other payments. The marketed surplus may be more, less or equal to the marketable surplus.

5. **Marketable Surplus($Y_5$):** The marketable surplus is that quantity of produce which can be made available to the nonfarm population of the country. It is a theoretical concept of surplus. The marketable surplus is the residual left with the producer farmer after meeting his requirement for family consumption, farm needs for needs for seeds and feed for cattle, payment to labour in kind. This may be expressed as follows:

$$MS = P - C$$

Where

MS = Marketable surplus,

P= Total production, and

C= Total requirements (family consumption, farm needs for needs for seeds and feed for cattle, payment to labour in kind)

## 4.5 PREPARATION OF INTERVIEW SCHEDULE

On the basis of the findings of pilot study a preliminary interview schedule was formed with the help of literature and by the assistance of Chairman of Advisory Committee. The interview schedule consisted of three major parts according to the specific objectives of the study.

## 4.6 PRE-TESTING OF INTERVIEW SCHEDULE

Pretesting or preliminary testing is the process of an advance testing of the study design after the schedule/questionnaire has been prepared. The object of pretesting is to detect the discrepancies that have emerged and to remove them after necessary modification in the schedule. It also helps to identify whether the questions are logically organized, the replies could properly have recorded in the space provided for or there is any scope for further improvement. After conducting pretesting appropriate changes and modification of the interview schedule have been made. The individuals who responded in pretesting have been excluded in the final sample selected for the study.

## 4.7 TECHNIQUES OF FIELD DATA COLLECTION

The respondents were personally interviewed during puja vacation and summer vacation. The items were asked in Bengali as well as English version in a simple term so that the members could understand easily. The entries were done in the schedule by student investigator himself at the time of interview.

## 4.8 STATISTICAL TOOLS FOR ANALYSIS AND INTERPRETATION OF DATA

The statistical methods used for analysis and interpretation of raw data were –

1. Mean
2. Standard deviation
3. Coefficient of Variance
4. Correlation of coefficient
5. Multiple regression analysis
6. Factor analysis
7. Path analysis
8. Canonical Correlation Analysis

### Mean

The mean is the arithmetic average and is the result obtained when the sum of the value of individual in the data is divided by the number of individuals in the data. Mean is simplest and relatively stable measure of central tendency. The mean reflects and is affected by every score in the distribution. We can work it out as follows

$$\text{Mean} = \overline{X} = \frac{\sum_{i=1}^{N} x_1}{N}$$

Where,

() = The symbol we use for mean (pronounced as x bar)

$\Sigma$ = Symbol for summation

$x_i$ = Value of the i$^{th}$ item X , I = 1, 2, ............................N

N = Total number of items.

Mean is the simplest measurement of central tendency and is a widely used measure. Its chief use consists in summarizing the essential features of a series and in enabling data to be compared. It is a relatively stable measure of central tendency. But it suffers from some limitations *viz.* it is unduly affected by extreme; it may not coincide with actual value of an item in a series, and it may lead to strong impressions, particularly when the item values are not given with the average. However, mean is better than other average, especially in economic and social studies where direct quantitative measurements are possible.

### Standard Deviation

Standard deviation is the most widely used measure of dispersion of a series and is commonly denoted by the symbol σ (pronounced as sigma) Standard deviation is the square root of the arithmetic mean of the square of the deviations, the deviations being measured from the arithmetic mean of distribution.

It is less affected by sampling errors and is more stable measure of dispersion. It is worked out as follows,

$$\text{Standard deviation } (\alpha) = \sqrt{\frac{\Sigma(x_i - x)^2}{n}}$$

**Coefficient of Variation**

A measure of variation which is independent of the unit of measurement is provided by Coefficient of variation. Being unit free, this is useful for computation of variability between different populations. The Coefficient of variation is standard deviation expressed as percentage of the mean and is measured by the formula.

$$\text{CV} = \frac{\text{Standard Deviation } (\sigma)}{\text{Mean}} \times 100$$

**Coefficient of Correlation**

When an increase or decrease in one variable is accompanied by an increase or decrease in other variable, the two are said to be correlated and the phenomenon is known as correlation. Correlation coefficient (r) is a measure of the relationship between two variables, which are at the interval or ratio level or measurement and are linearly related. A Karl Pearson's coefficient of correlation also known as product moment 'r' is computed by the formula.

$$r_{xy} = \frac{N\Sigma xy - (\Sigma x)(\Sigma y)}{\sqrt{[N\Sigma^2 - (\Sigma x)^2][N\Sigma y^2 - (\Sigma y)^2]}}$$

Where,

x and y = Original scores in variables x and y

N = Number of paired scores

Σxy = Each x multiplied by its corresponding y, then summed

Σx = Sum of x scores

$\Sigma x^2$ = Each x squared, then summed

$(\Sigma x)^2$ = Sum of x scores, squared

Σy = Sum of y scores

$\Sigma y^2$ = each y squared, then summed

$(\Sigma y)^2$ = Sum of y scores, squared

This coefficient assumes the following;

That there is linear relationship between the two variables;

That the two variables are causally related which means that one of the variable is independent and other is dependent and; A large number of independent causes are operating in both variables so as to produce a normal distribution.

The value of 'r' lies between +1 to -1. Positive values of r indicate that positive correlation between the two variables (i.e. changes in both variables take place in the same direction), whereas negative values of 'r' indicate negative correlation i.e. changes in the two variables taking place in opposite direction. A zero value of 'r' indicates that there is no association between the two variables. When r (+) 1, it indicates perfect positive correlation and when it is (-) 1, it indicates perfect negative correlation, meaning thereby that variations in independent variable (x) explain 100 per cent of the variations in the dependent variable (y). We can also say that for a unit change in independent variable, if there happens to be constant change in the dependent variable in the same direction, the correlation will be termed as perfect positive. But if such change occurs in the opposite direction, the correlation will be termed as perfect negative. The value of 'r' nearer to +1 or -1 indicates high degree of correlation between the two variables.

## Regression

The correlation coefficient only expresses association and by itself tells nothing about the causal relationships of the variables.

Thus, purely from the knowledge that two variables x and y are correlated, we cannot say whether variation in x is the cause or the results from mutual dependence of the two variables or from common causes affecting both of them. Similarly, the mere existence of a high value of correlation coefficient is not necessarily of an underlying relationship between the two variables.

The underlying relation between y and x in a bivariate population can be expressed in the form of a mathematical equation known as regression equation and is said to represent the regression of the variable y on the variable x. (Panse and Sukhatme, 1967)

If y is the dependent variable and x is the independent variable, then the linear regression equation can be written as

$$y = a + bx$$

The values of a and b can be obtained by the method of least squares which consists of minimizing the expression

$\Sigma(y_i - a - bx_i)^2$ with respect to a and b.

The values of a and b are

$$a = y - bx$$

$$b = \frac{\Sigma xy - \frac{(\Sigma xi)(\Sigma yi)}{n}}{\Sigma xi^2 - \frac{(\Sigma x)^2}{n}}$$

The regression equation can now be written as

$$y = y - bx + bx$$

$$y - y = b(x - x)$$

Where b is the regression coefficient

## Stepwise Multiple Regression

Stepwise regression is a variation of multiple regressions which provides a means of choosing independent variables that yield the best prediction possible with the fewest independent variables. It permits the user to solve a sequence of one or more multiple linear regression problems by stepwise application of the least square method.

At each step in the analysis, a variable is added or removed which results in the greatest production in the error sum of squares (Burroughs Corporation, 1975).

According to Drapper and Smith (1981), the method of stepwise multiple regression analysis is to insert variables in turn until the regression equation in satisfactory. The order of insertion is determined by suing the partial correlation coefficient as a measure of the importance of variables not yet in the equation.

The program, according to Burroughs Corporation (1975), first forms a correlation matrix, finds the best predictor (the independent variable having the highest correlation with criterion variable) and performs a regression analysis with this predictor. Then, the second best predictor (independent), and so on. At any given step, the group of predictors being used

is not necessarily the best group of that size (i.e. the particular group of independent variables does not necessarily have the highest multiple correlation with the criterion that any group of this size does). Rather, this group contains the variables that have the highest individual correlation with the criterion.

Significance of variable that is being considered for entrance into the regression equation is measured by the F-statistic. If F is too small (less than F 'include'), the variable is not added to the regression equation. Include statement establishes the minimum value of the F-statistic required for the inclusion of a variable in the regression equation. In the example which follows, the F-value for inclusion was 0.01.

Significance of variables already in the regression equation may change as new variables are entered. This significance of the variables currently in the equation is also measured by the F-statistic. If F is too small (less than F 'delete'), the variable is not added to the equation. Delete establishes the value of the F-statistic below which the variable is deleted from the regression equation. Here, the F-value for deletion was 0.005.

The 'tolerance' level specified is used as control of degeneracy occurs when a variable entered into the equation is a linear combination of variables already present in the equation. Tolerance statement establishes the maximum value a pivoted element may attain while still allowing its associated variable to be brought into equation. A variable is not brought into the regression equation if its associated pivoted element is below the specified tolerance level, which was 0.001 in the present example.

## Factor Analysis

Factor analysis is a very useful and popular method of multivariate research technique, mostly used in social and behavioural sciences. This technique is applicable when there is a systematic interdependence among a set of observed or manifest variables, and the researcher is interested in finding out something more fundamental or latent which creates this communality (commonness).

For example, we may have data on farmers' education, occupation, land, house, farm power, material possession, social participation etc. and want to infer from these some factor relating to social status, which shall summarize the communality of all the variables.

According to Kothari (1996), Factor analysis seeks to resolve a large set of measured variables in terms of relatively few categories, known as factors. This technique allows the researcher to group variables into factors (based on correlation between variables), and the factors so derived may be treated as new variables (often termed as latent variables) and their grouped into the factor. The meaning and name of such new variable is subjectively determined by the researcher.

Since the factors happen to be linear combinations of data, the coordinates of each observation or variables is measured to obtain what are called factor loadings. Such factors loadings represent the correlation between the particular variable and the factor, and are usually placed in a matrix of correlations between the variable and the factors.

## Concepts Used in Factor Analysis

Some important concepts used in factor analysis are explained, following Kothari (1996).

**Factor:** A factor is an underlying dimension that accounts for several observed variables. Factor is a hypothetical construct or classification. There may be one or more factors, depending upon the nature of the study and the number of variables involved in it.

**Factor loadings**: Factor loadings are those values which explain how closely the variables are related to each one of the factors discovered. Factor loadings work as key to understanding what the factors mean. It is the absolute size (rather the signs, plus or minus) of the loadings that is important in the interpretation of a factor.

**Communality ($h^2$)** Communality, represented by $h^2$, shows how much of each variable is accounted for by the underlying factor taken together. A high value of communality means that not much of the variable is left over after whatever the factors represent is taken into consideration. Variables are taken into account. There are some steps in factor analysis:

We have to collect data then we have to work out the correlation between the variables. It is to explore the possibility of data reduction i.e. initial steps of factor are to be explored. The common method of extraction of factors is Principle Component Analysis (P.C.A).

## Principal Component Analysis

There are several methods of factor analysis. The method of Principal Component Analysis which is widely used is discussed here.

The principal component analysis extracts m-eigenvectors (principal component axes) and corresponding m-eigenvalues (the variance measured along the eigenvector), from m x m symmetrical matrix of correlation.

The eigenvectors obtained from this principal component analysis are all orthogonal (i.e. inter-column correlations are near zero). The eigenvalues account for all of the original data variances in decreasing order such that each has variance or eigenvalue less than the previous ones. The total of the eigenvalues

$$(\lambda_1 + \lambda_2 + \dots \dots \dots + \lambda_m)$$

which is the same as the sum of the variances constituting the diagonal or trace of the correlation matrix before transformation. The principal components are then converted into factors by multiplying each element of the principal components or eigenvectors (v) by the square – root of the corresponding eigenvalues ($\lambda^{1/2}.V$). Factors, thus, besides the direction also represent the variances.

The analysis calls for the selection of a minimum number of meaningful and useful factors, considerably fewer in number than the original variables, which will account for most of the variances in the data set and therefore, convey the same information. Various criteria for selection of suitable factors are available. Kaiser (1958) and others have recommended retaining all those eigenvalues, which have values more than one.

Next step is to remove the noise imposed by (m - p) unnecessary axes. To accomplish this, p-orthogonal reference axes or factors are routed about the origin to positions such that the variance of the loading from each variable onto each factor axis is either extreme (±1) or zero. This maximization of the range of the loadings was performed by using Kaiser's Varimax criterion. Scanning through each factor column for large absolute values in the varimax matrix will reveal a few variables with significantly high loadings and many

**Eigenvalue (or latent root):** The sum of squared values of factor loadings relating to a factor is referred to as Eigen value or latent root. Eigenvalue indicates the relative importance of each factor in accounting for the particular set of variables being analysed.

**Rotation:** Rotation reveals different structures in the data and provides meaning to the results of factor analysis. There are different types of rotations such as orthogonal rotations, oblique rotations, varimax rotation etc. One has to select a rotation appropriate to the study. For the present study varimax rotation has been used.

Factor Analysis is used:

To reduce the dimensionality of large number of variables to a fewer number of factors.

To confirm the hypothesized factor structure by way of testing of hypothesis about the structure of variables in terms of expected number of significant factor loading.

Hence in factor analysis specific and error variables are excluded and only the common others with insignificantly loadings. The column showing communality ($\Sigma h^2_j$) is the total amount of variance of each variable retained in the factors, and is computed by summing the squares of the elements of the factors in each row of the varimax matrix. Fairly high communality of each variable implies the appropriateness of the model adopted, for the study. The last step involved interpretation of the factors.

## Canonical Correlation Analysis

Canonical correlation analysis is used to identify and measure the associations among two sets of variables. Canonical correlation is appropriate in the same situations where multiple regression would be, but where are there are multiple inter-correlated outcome variables. Canonical correlation analysis determines a set of canonical variates, orthogonal linear combinations of the variables within each set that best explain the variability both within and between sets. Canonical correlation analysis can be defined as the problem of finding two sets of basis vectors, one for x and the other for y, such that the correlations between the projections of the variables onto these basis vectors are mutually maximized. Multiple regression analysis is a multivariate technique which can predict the value of a single (metric) dependent variable from a linear function of a set of independent variables. For some research problems, however, interest may not center on a single dependent variable; rather, the researcher may be interested in relationships between sets of multiple dependent and multiple independent variables. **Canonical correlation analysis** is a multivariate statistical model that facilitates the study of interrelationships among sets of multiple dependent variables and multiple independent variables. Whereas multiple regression predicts a single dependent variable from a set of multiple independent variables, canonical correlation simultaneously predicts multiple dependent variables from multiple independent variables. Canonical correlation places the fewest restrictions on the types of data on which it operates. Because the other techniques impose more rigid restrictions, it is generally believed that the information obtained from them is of higher quality and may be presented in more interpretable manner. For this reason, many researchers view canonical correlation as a last-ditch effort, to be used when all other higher-level techniques have been exhausted. But in situations with multiple dependent and independent variables, canonical correlation is the most appropriate and powerful multivariate technique. It has gain acceptance in many fields and represents a useful tool for multivariate analysis, particularly as interest has spread to considering multiple dependent variables.

To clarify further the nature of canonical correlation, let us consider an extension of a simple example of multiple regression analysis. Assume that a survey was conducted to understand the relationships between family size and income as predictors of the number of credit cards a family would hold. Such a problem would involve examining the relationship between two independent variables and a single dependent variable.

Suppose the researcher then became interested in the broader concept of credit usage. To measure credit usage, the researcher considered not only the number of credit cards held by the family but also the family's average monthly dollar charges on all credit cards. These two measures were felt to give a much better perspective on a family's credit card usage. Readers interested in the approach of using multiple indicators to represent a concept are referred to discussions of Factor Analysis and Structural Equation Modeling. The problem now involves predicting two dependent measures simultaneously (number of credit cards and average dollar charges). Multiple regression is capable of handling only a single dependent variable. Multivariate analysis of variance could be used, but only if all of the independent variables were non-metric, which is not the case in this problem. Canonical correlation represents the only technique available for examining the relationship with multiple dependent variables. By using canonical correlation analysis, the researcher creates a composite measure of credit usage that consists of both dependent variables, rather than having to compute a separate regression equation for each of the dependent variables. The result of applying canonical correlation is a measure of the strength of the relationship between two sets of multiple variables (canonical variates). The measure of the strength of the relationship between the two variates is expressed as a canonical correlation coefficient (Rc). The researcher now has two results of interest: the canonical variates representing the optimal linear combinations of dependent and independent variables; and the canonical correlation representing the Canonical correlation analysis is the most generalized member of the family of multivariate statistical techniques. It is directly related to several dependence methods. Similar to regression, canonical correlation's goal is to quantify the strength of the relationship, in this case between the two sets of variables (independent and dependent). It corresponds to factor analysis in the creation of composites of variables. It also resembles discriminant analysis in its ability to determine independent dimensions (similar to discriminant functions) for each variable set, in this situation with the objective of producing the maximum correlation between the dimensions. Thus, canonical correlation identifies the optimum structure or dimensionality of each variable set that maximizes the relationship between independent and dependent variable sets. Canonical correlation analysis deals with the association between the composites of sets of multiple dependent and independent variables. In doing so, it develops a number of independent **canonical functions** that maximize the correlation between the **linear composites**, also known as **canonical variates**, which are sets of dependent and independent variables. Each canonical function is actually based on the correlation between two canonical variates, one variate for the dependent variables and one for the independent variables. Another unique feature of canonical correlation is that the variates are derived to maximize their correlation. Moreover, canonical relationship between them. Correlation does not stop with the derivation of a single relationship between the sets of variables. Instead, a number of canonical functions (pairs of canonical variates) may be derived.

## Path Analysis

The objective of doing Path Analysis is to get a clear picture of the direct and indirect effects of the independent variables on the dependent variable. Variables, through which substantial indirect effects are channelled, are also found out.

Singh and Chaudhary (1977), defined path coefficient as the ratio of the standard deviation of the effect due to a given cause to the total standard deviation of the effect i.e. if Y is the effect and x1 is the cause, the path coefficient for the path from cause x1 to the effect Y is $\sigma x1/\sigma y$.

It is advisable to do path analysis with only those variables which have significant effects on the dependent variable. This may be done by restoring to multiple regression analysis, and selecting those independent variables whose partial b value are significant. This shall enhance clarity of the path analysis.

5

CHAPTER

# RESULTS AND DISCUSSION

**Table No 5.1. General distribution of statistical character of variables: Dependent and Independent variables.**

| Variables | Range | | Mean | Std. Deviation | CV (%) |
|---|---|---|---|---|---|
| | Minimum | Maximum | | | |
| 1. Age ($X_1$) | 21.00 | 54.00 | 38.70 | 9.57 | 24.742 |
| 2. Education ($X_2$) | 1.00 | 10.00 | 4.80 | 2.38 | 49.583 |
| 3. Family size ($X_3$) | 2.00 | 19.00 | 6.50 | 4.18 | 64.358 |
| 4. Family education ($X_4$) | 1.00 | 10.00 | 4.55 | 2.01 | 44.174 |
| 5. Area under pulse cultivation (Bigha) ($X_5$) | 0.23 | 5.00 | 0.98 | 0.74 | 75.508 |
| 6. Farm size and technology adoption (Bigha) ($X_6$) | 1.00 | 21.00 | 9.02 | 5.01 | 55.532 |
| 7. No of crop diversity ($X_7$) | 0.09 | 2.33 | 0.79 | 0.48 | 60.672 |
| 8. Income (Rs /per capita/annum) ($X_8$) | 9102.00 | 84061.00 | 46414.82 | 20744.37 | 44.693 |
| 9. Risk Orientation ($X_9$) | 2.66 | 8.83 | 5.45 | 1.17 | 21.617 |
| 10. Scientific Orientation ($X_{10}$) | 2.00 | 8.00 | 5.42 | 1.20 | 22.160 |
| 11. Planning orientation ($X_{11}$) | 3.00 | 8.33 | 5.61 | 1.16 | 20.736 |
| 12. Production orientation ($X_{12}$) | 2.66 | 8.33 | 5.50 | 1.18 | 21.576 |
| 13. Market Orientation ($X_{13}$) | 2.00 | 6.66 | 4.54 | 0.94 | 20.875 |
| 14. Farmers attitude towards Pulse crop cultivation ($X_{14}$) | 2.30 | 7.00 | 4.46 | 0.95 | 21.456 |
| 15. Knowledge level of farmer towards cultivation of pulse crop ($X_{15}$) | 6.00 | 13.00 | 8.98 | 2.20 | 24.528 |
| 16. Knowledge about insecticides ($X_{16}$) | 1.00 | 2.00 | 1.48 | 0.20 | 13.948 |
| 17. Knowledge about fungicide ($X_{17}$) | 1.00 | 2.00 | 1.51 | 0.20 | 13.667 |

| Variables | Range | | Mean | Std. Deviation | CV (%) |
|---|---|---|---|---|---|
| | Minimum | Maximum | | | |
| 18. Knowledge about weed control ($X_{18}$) | 1.00 | 2.00 | 1.45 | 0.34 | 23.840 |
| 19. Knowledge about IPM practice ($X_{19}$) | 1.25 | 1.87 | 1.50 | 0.13 | 9.039 |
| 20. Farmers attitude towards IPM programme ($X_{20}$) | 3.50 | 7.14 | 5.38 | 0.74 | 13.931 |
| 21. Attitude towards adoption ($X_{21}$) | 3.60 | 8.20 | 5.53 | 0.85 | 15.441 |
| 22. Attitude towards discontinuous ($X_{22}$) | 5.66 | 9.88 | 7.88 | 1.00 | 12.743 |
| 23. Attitude towards rejection ($X_{23}$) | 5.73 | 9.00 | 6.99 | 0.52 | 7.451 |
| 24. Productivity or yield (kg/Bigha) ($X_{24}$) | 251.00 | 416.00 | 337.12 | 47.95 | 14.225 |
| 25. Gross return (Rs/ Bigha) ($X_{25}$) | 2239.50 | 38872.00 | 9842.24 | 6654.10 | 67.608 |
| 26. Utilization of cosmopolite sources of information ($X_{26}$) | 2.00 | 3.13 | 2.51 | 0.21 | 8.718 |
| 27. Training received ($X_{27}$) | 0.00 | 15.00 | 4.03 | 2.82 | 70.092 |
| 28. Socialization of pulse enterprise ($Y_1$) | 5.81 | 542.50 | 100.79 | 85.83 | 85.155 |
| 29. Socialization through replacement level ($Y_2$) | 1.21 | 75.00 | 16.55 | 17.16 | 103.664 |
| 30. Socialization through reinvention ($Y_3$) | 33.33 | 400.00 | 126.22 | 75.14 | 59.532 |
| 31. Marketed surplus ($Y_4$) | 130.52 | 328.80 | 213.63 | 44.38 | 20.776 |
| 32. Marketable surplus ($Y_5$) | 154.80 | 328.64 | 235.68 | 38.65 | 16.401 |

Table No 6.1. Presents general distribution of the statistical character of variables (dependent and independents).

From the table 6.1 it is very clear that age ($X_1$) is having the ranges (Min-Max) between 21 years to 54 years, and the mean value of age is 38.70, with standard deviation is 9.57, for the total distribution taken for the study. The coefficient of variation (C.V) for total distribution is 24.74 per cent. It appears from the CV (%) that the distribution of variable stands highly consistent.

The ranges (Min-Max) of the variable, education ($X_2$) between 1 to 10, and the mean value is 4.80, with standard deviation is 2.38, for the total distribution taken for the study. The coefficient of variation (C.V) for total distribution is 49.58 per cent. It appears from the CV (%) that the distribution of variable stands highly consistent.

The ranges (Min-Max) of the variable, family size ($X_3$) between 2.0 to 19.00, and the mean value is 6.50, with standard deviation is 4.18, for the total distribution taken for the study. The coefficient of variation (C.V) for total distribution is 64.35 per cent. It appears from the CV (%) that the distribution of variable stands Moderately consistent.

The ranges (Min-Max) of the variable, family education ($X_4$) between 1 to 10, and the mean value is 4.55, with standard deviation is 2.014, for the total distribution taken for the study. The coefficient of variation (C.V) for total distribution is 44.17 per cent. It appears from the CV (%) that the distribution of variable stands highly consistent.

The ranges (Min-Max) of the variable, area under pulse cultivation (Bigha) ($X_5$) between 0.23 to 5.00, and the mean value is 0.98, with standard deviation is 0.74, for the total distribution taken for the study. The coefficient of variation (C.V) for total distribution is 75.50 per cent. It appears from the CV (%) that the distribution of variable stands moderately consistent.

The ranges (Min-Max) of the variable, farm size and technology adoption (Bigha) ($X_6$) between 1.00 to 21.00, and the mean value is 9.02, with standard deviation is 5.01, for the total distribution taken for the study. The coefficient of variation (C.V) for total distribution is 55.53 per cent. It appears from the CV (%) that the distribution of variable stands moderately consistent.

The ranges (Min-Max) of the variable, number of crop diversity ($X_7$) between .095 to 2.33, and the mean value is 0.795, with standard deviation is 0.482, for the total distribution taken for the study. The coefficient of variation (C.V) for total distribution is 60.67 per cent. It appears from the CV (%) that the distribution of variable stands moderately consistent.

The ranges (Min-Max) of the variable, income (rs /per capita/annum) ($X_8$) between 9102.00 to 84061.000, and the mean value is 46414.82, with standard deviation is 20744.37, for the total distribution taken for the study. The coefficient of variation (C.V) for total distribution is 44.69 per cent. It appears from the CV (%) that the distribution of variable stands highly consistent.

The ranges (Min-Max) of the variable, risk Orientation ($X_9$) between 2.66 to 8.83, and the mean value is 5.45, with standard deviation is 1.178, for the total distribution taken for the study. The coefficient of variation (C.V) for total distribution is 21.61 per cent. It appears from the CV (%) that the distribution of variable stands highly consistent.

The ranges (Min-Max) of the variable, scientific Orientation ($X_{10}$) between 2.00 to 8.00, and the mean value is 5.42, with standard deviation is 1.20, for the total distribution taken for the study. The coefficient of variation (C.V) for total distribution is 22.16 per cent. It appears from the CV (%) that the distribution of variable stands highly consistent.

The ranges (Min-Max) of the variable, planning orientation ($X_{11}$) between 3.00 to 8.33, and the mean value is 5.61, with standard deviation is 1.16, for the total distribution taken for the study. The coefficient of variation (C.V) for total distribution is 20.73 per cent. It appears from the CV (%) that the distribution of variable stands highly consistent.

The ranges (Min-Max) of the variable, management orientation ($X_{12}$) between 2.66 to 8.33, and the mean value is 5.50, with standard deviation is 1.18, for the total distribution taken for the study. The coefficient of variation (C.V) for total distribution is 21.57 per cent. It appears from the C.V (%) that the distribution of variable stands highly consistent.

The ranges (Min-Max) of the variable, market orientation ($X_{13}$) between 2.00 to 6.66, and the mean value is 4.54, with standard deviation is 0.948, for the total distribution taken for the study. The coefficient of variation (C.V) for total distribution is 20.87 per cent. It appears from the CV (%) that the distribution of variable stands highly consistent.

The ranges (Min-Max) of the variable, farmer's attitude towards pulse crop cultivation ($X_{14}$) between 2.30 to 7.00, and the mean value is 4.46, with standard deviation is .957, for the total distribution taken for the study. The coefficient of variation (C.V) for total distribution is 21.45 per cent. It appears from the CV (%) that the distribution of variable stands highly consistent.

The ranges (Min-Max) of the variable, knowledge level of farmer towards cultivation of pulse crop ($X_{15}$) between 6.00 to 13.00, and the mean value is 8.98, with standard deviation is 2.20, for the total distribution taken for the study. The coefficient of variation (C.V) for total distribution is 24.52 per cent. It appears from the CV (%) that the distribution of variable stands highly consistent.

The ranges (Min-Max) of the variable, knowledge about insecticides ($X_{16}$) between 1.00 to 2.00, and the mean value is 1.4, with standard deviation is 0.20, for the total distribution taken for the study. The coefficient of variation (C.V) for total distribution is 13.94 per cent. It appears from the CV (%) that the distribution of variable stands highly consistent.

The ranges (Min-Max) of the variable, knowledge about fungicide ($X_{17}$) between 1.00 to 2.00, and the mean value is 1.51, with standard deviation is 0.20, for the total distribution taken for the study. The coefficient of variation (C.V) for total distribution is 13.66 per cent. It appears from the CV (%) that the distribution of variable stands highly consistent.

The ranges (Min-Max) of the variable, knowledge about weed control ($X_{18}$) between 1.00 to 2.00, and the mean value is 1.45, with standard deviation is 0.34, for the total distribution taken for the study. The coefficient of variation (C.V) for total distribution is 23.84 per cent. It appears from the CV (%) that the distribution of variable stands highly consistent.

The ranges (Min-Max) of the variable, knowledge about IPM practice ($X_{19}$) between 1.25 to 1.87, and the mean value is 1.50, with standard deviation is .136, for the total distribution taken for the study. The coefficient of variation (C.V) for total distribution is 9.03 per cent. It appears from the CV (%) that the distribution of variable stands highly consistent.

The ranges (Min-Max) of the variable, farmer's attitude towards IPM programme ($X_{20}$) between 3.50 to 7.14, and the mean value is 5.38, with standard deviation is .749, for the total distribution taken for the study. The coefficient of variation (C.V) for total distribution is 13.93 per cent. It appears from the CV (%) that the distribution of variable stands highly consistent.

The ranges (Min-Max) of the variable, attitude towards adoption ($X_{21}$) between 3.60 to 8.20, and the mean value is 5.53, with standard deviation is 0.854, for the total distribution taken for the study. The coefficient of variation (C.V) for total distribution is 15.44 per cent. It appears from the CV (%) that the distribution of variable stands highly consistent.

The ranges (Min-Max) of the variable, attitude towards discontinues ($X_{22}$) between 5.66 to 9.88, and the mean value is 7.88, with standard deviation is 1.00, for the total distribution taken for the study. The coefficient of variation (C.V) for total distribution is 12.74 per cent. It appears from the CV (%) that the distribution of variable stands highly consistent.

The ranges (Min-Max) of the variable, attitude towards rejection ($X_{23}$) between 5.73 to 9.00, and the mean value is 6.99, with standard deviation is 0.52, for the total distribution taken for the study. The coefficient of variation (C.V) for total distribution is 7.45 per cent. It appears from the CV (%) that the distribution of variable stands highly consistent.

The ranges (Min-Max) of the variable, productivity or yield (kg/Bigha) ($X_{24}$) between 251.00 to 416.00, and the mean value is 337.12, with standard deviation is 47.95, for the total distribution taken for the study. The coefficient of variation (C.V) for total distribution is 14.22 per cent. It appears from the CV (%) that the distribution of variable stands highly consistent.

The ranges (Min-Max) of the variable, gross return (Rs/ Bigha) ($X_{25}$) between 2239.50 to 38872.00 and the mean value is 9842.24, with standard deviation is 6654.101, for the total distribution taken for the study. The coefficient of variation (C.V) for total distribution is 67.60 per cent. It appears from the CV (%) that the distribution of variable stands moderately consistent.

The ranges (Min-Max) of the variable utilization of cosmopolite sources of information($X_{26}$) between 2.00 to 3.13 and the mean value is 2.51, with standard deviation is 0.21, for the total distribution taken for the study. The coefficient of variation (C.V) for total distribution is 8.71 per cent. It appears from the CV (%) that the distribution of variable stands highly consistent.

The ranges (Min-Max) of the variable training received ($X_{27}$) between 0.00 to 15.00 and the mean value is 4.03, with standard deviation is 2.82, for the total distribution taken for the study. The coefficient of variation (C.V) for total distribution is 70.09 per cent. It appears from the CV (%) that the distribution of variable stands moderately consistent.

The ranges (Min-Max) of the variable socialization of pulse enterprise ($Y_1$) between 5.81 to 542.50 and the mean value is 100.79, with standard deviation is 85.83, for the total distribution taken for the study. The coefficient of variation (C.V) for total distribution is 85.15 per cent. It appears from the CV (%) that the distribution of variable stands moderately consistent.

The ranges (Min-Max) of the variable socialization through replacement level ($Y_2$) between 1.21 to 75.00 and the mean value is 16.55, with standard deviation is 17.16, for the total distribution taken for the study. The coefficient of variation (C.V) for total distribution is 103.66 per cent. It appears from the CV (%) that the distribution of variable stands moderately consistent.

The ranges (Min-Max) of the variable socialization through reinvention level ($Y_3$) between 33.33 to 400.00 and the mean value is 126.22, with standard deviation is 75.14, for the total distribution taken for the study. The coefficient of variation

(C.V) for total distribution is 59.53 per cent. It appears from the CV (%) that the distribution of variable stands moderately consistent.

The ranges (Min-Max) of the variable marketed surplus ($Y_4$) between 130.52 to 328.80 and the mean value is 213.63, with standard deviation is 44.38 for the total distribution taken for the study. The coefficient of variation (C.V) for total distribution is 20.77 per cent. It appears from the CV (%) that the distribution of variable stands highly consistent.

The ranges (Min-Max) of the variable marketable surplus ($Y_5$)) between 154.8 to 328.64 and the mean value is 235.68, with standard deviation is 38.65, for the total distribution taken for the study. The coefficient of variation (C.V) for total distribution is 16.40 per cent. It appears from the CV (%) that the distribution of variable stands highly consistent.

**Table No 6.2. Correlation coefficient of Socialization of pulse enterprise ($Y_1$) with 27 independent variables**

| Variables | r- value | Sig. (2-tailed) |
|---|---|---|
| 1. Age ($X_1$) | -0.043 | 0.599 |
| 2. Education ($X_2$) | 0.031 | 0.705 |
| 3. Family size ($X_3$) | .238** | 0.003 |
| 4. Family education ($X_4$) | -0.023 | 0.779 |
| 5. Area under pulse cultivation (Bigha) ($X_5$) | .830** | 0.000 |
| 6. Farm size and technology adoption (Bigha) ($X_6$) | 0.049 | 0.553 |
| 7. No of crop diversity ($X_7$) | -0.094 | 0.255 |
| 8. Income (Rs /per capita/annum) ($X_8$) | -0.042 | 0.614 |
| 9. Risk Orientation ($X_9$) | -0.064 | 0.435 |
| 10. Scientific Orientation ($X_{10}$) | 0.035 | 0.670 |
| 11. Planning orientation ($X_{11}$) | 0.016 | 0.847 |
| 12. Production orientation ($X_{12}$) | -0.014 | 0.864 |
| 13. Market Orientation ($X_{13}$) | 0.090 | 0.273 |
| 14. Farmers attitude towards Pulse crop cultivation ($X_{14}$) | .523** | 0.000 |
| 15. Knowledge level of farmer towards cultivation of pulse crop ($X_{15}$) | .391** | 0.000 |
| 16. Knowledge about insecticides ($X_{16}$) | -0.008 | 0.925 |
| 17. Knowledge about fungicide ($X_{17}$) | 0.066 | 0.424 |
| 18. Knowledge about weed control ($X_{18}$) | 0.015 | 0.858 |
| 19. Knowledge about IPM practice ($X_{19}$) | 0.050 | 0.540 |
| 20. Farmers attitude towards IPM programme ($X_{20}$) | 0.056 | 0.496 |
| 21. Attitude towards adoption ($X_{21}$) | 0.049 | 0.555 |
| 22. Attitude towards discontinuous ($X_{22}$) | 0.084 | 0.307 |
| 23. Attitude towards rejection ($X_{23}$) | -0.028 | 0.737 |

| Variables | r- value | Sig. (2-tailed) |
|---|---|---|
| 24. Productivity or yield (kg/Bigha) ($X_{24}$) | -.202* | 0.013 |
| 25. Gross return (Rs/ Bigha) ($X_{25}$) | .834** | 0.000 |
| 26. Utilization of cosmopolite sources of information($X_{26}$) | -0.065 | 0.427 |
| 27. Training received ($X_{27}$) | .378** | 0.000 |

*Correlation is significant at 5 % level, **Correlation is significant at 1 % level.

## Results

Table 6.2. Presents the correlation coefficient of consequent variable, socialization of pulse enterprise ($Y_1$) with 27 independent variables.

It has been found that variables viz. family size ($X_3$), area under pulse cultivation (bigha) ($X_5$), farmer's attitude towards pulse crop cultivation ($x_{14}$), knowledge level of farmer towards cultivation of pulse crop ($X_{15}$), gross return (rs/ bigha) ($X_{25}$), training received ($X_{27}$) have recorded the positive and productivity or yield (kg/ bigha) ($X_{24}$) has recorded significant but negative correlation with socialization of pulse enterprise.

## Revelation

Result evinces that the socialization of pulse enterprise ($Y_1$) has scaled up for those having higher size of holding and these also been elicited by the farmer having proper attitude and adequate knowledge for the cultivation viz for socialization of pulse enterprise ($Y_1$).

Socialization of pulse enterprise ($Y_1$) also has helped the scaling up of gross return became of training received and exposure met subsequently however its interestingly to know that pulse productivity has been better for those having less of exposure to formal socialization programme organization formal institution as well as organization.

**Model No. 6.1. Model of Correlation coefficient of Socialization of pulse enterprise ($Y_1$) with 27 independent variables**

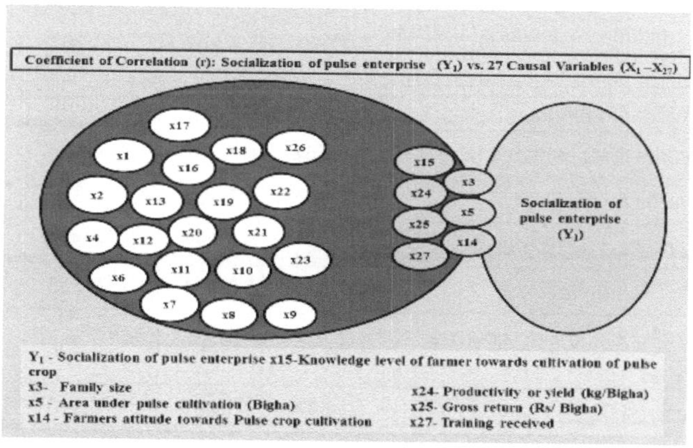

**Table No 6.3 Coefficients of Multiple regression analysis of Socialization of pulse enterprise ($Y_1$) with 27 causal variables**

| Variables | Unstandardized Coefficients | | Standardized Coefficients | T | Sig. |
|---|---|---|---|---|---|
| | B | Std. Error | Beta | | |
| 1. Age ($X_1$) | -0.113 | 0.405 | -0.013 | -0.278 | 0.781 |
| 2. Education ($X_2$) | 0.634 | 1.714 | 0.018 | 0.370 | 0.712 |
| 3. Family size ($X_3$) | -0.896 | 1.013 | -0.044 | -0.885 | 0.378 |
| 4. Family education ($X_4$) | 0.572 | 2.023 | 0.013 | 0.283 | 0.778 |
| 5. Area under pulse cultivation (Bigha) ($X_5$) | 41.217 | 12.837 | 0.359 | 3.211 | 0.002 |
| 6. Farm size and technology adoption (Bigha) ($X_6$) | -0.250 | 1.086 | -0.015 | -0.230 | 0.818 |
| 7. No of crop diversity ($X_7$) | -10.006 | 11.067 | -0.056 | -0.904 | 0.368 |
| 8. Income (Rs /per capita/annum) ($X_8$) | 6.994-005 | 0.000 | 0.017 | 0.364 | 0.717 |
| 9. Risk Orientation ($X_9$) | 1.974 | 3.335 | 0.027 | 0.592 | 0.555 |
| 10. Scientific Orientation ($X_{10}$) | -4.508 | 3.412 | -0.063 | -1.321 | 0.189 |
| 11. Planning orientation ($X_{11}$) | 2.325 | 3.388 | 0.032 | 0.686 | 0.494 |
| 12. Production orientation ($X_{12}$) | -1.113 | 3.206 | -0.015 | -0.347 | 0.729 |
| 13. Market Orientation ($X_{13}$) | 4.945 | 4.209 | 0.055 | 1.175 | 0.242 |
| 14. Farmers attitude towards Pulse crop cultivation ($X_{14}$) | 10.401 | 4.638 | 0.116 | 2.242 | 0.027 |
| 15. Knowledge level of farmer towards cultivation of pulse crop ($X_{15}$) | 3.902 | 2.039 | 0.100 | 1.913 | 0.058 |
| 16. Knowledge about insecticides ($X_{16}$) | -2.325 | 18.476 | -0.006 | -0.126 | 0.900 |
| 17. Knowledge about fungicide ($X_{17}$) | -17.517 | 19.005 | -0.042 | -0.922 | 0.358 |
| 18. Knowledge about weed control ($X_{18}$) | 5.824 | 11.840 | 0.024 | 0.492 | 0.624 |
| 19. Knowledge about IPM practice ($X_{19}$) | -13.185 | 27.920 | -0.021 | -0.472 | 0.638 |
| 20. Farmers attitude towards IPM programme ($X_{20}$) | 0.392 | 5.207 | 0.003 | 0.075 | 0.940 |
| 21. Attitude towards adoption ($X_{21}$) | 0.651 | 4.633 | 0.006 | 0.141 | 0.888 |
| 22. Attitude towards discontinuous ($X_{22}$) | -1.762 | 3.987 | -0.021 | -0.442 | 0.659 |
| 23. Attitude towards rejection ($X_{23}$) | 3.667 | 7.465 | 0.022 | 0.491 | 0.624 |
| 24. Productivity or yield (kg/Bigha) ($X_{24}$) | -0.229 | 0.086 | -0.128 | -2.662 | 0.009 |
| 25. Gross return (Rs/ Bigha) ($X_{25}$) | 0.005 | 0.001 | 0.395 | 3.650 | 0.000 |
| 26. Utilization of cosmopolite sources of information($X_{26}$) | 2.240 | 17.823 | 0.006 | .126 | 0.900 |
| 27. Training received ($X_{27}$) | 3.577 | 1.397 | 0.118 | 2.561 | 0.012 |

**Table No 6.4. Multiple regression analysis of Socialization of pulse enterprise ($Y_1$) with 27 causal variables (Model Summary)**

| Model | R | R Square | Adjusted R Square | Std. Error of the Estimate | Change Statistics | | | | |
|---|---|---|---|---|---|---|---|---|---|
| | | | | | R Square Change | F Change | df1 | df2 | Sig. F Change |
| 1 | 0.887[a] | 0.787 | 0.740 | 43.730079 | 0.787 | 16.741 | 27 | 122 | 0.000 |

## Results

Table No 6.3 Presents the multiple regression analysis which reflects the functional efficacy of the correlation through 'beta' value and respective 't' values of the casual variables on the consequent variables i.e. socialization of pulse enterprise ($Y_1$).

Table No 6.4 Presents The $R^2$ value 0.787 being, it's is to conclude that even combination of all 27 variables so far, 78.7 % of variability embedded with the consequent variable has been explained.

**Table No 6.5. Stepwise regression analysis of Socialization of pulse enterprise ($Y_1$) with 27 causal variables**

| Model | | Unstandardized Coefficients | Standardized Coefficients | T | Sig. | |
|---|---|---|---|---|---|---|
| | | B | Std. Error | Beta | | |
| 1 | (Constant) | -5.083 | 6.944 | | -0.732 | 0.465 |
| | $X_{25}$ | 0.011 | 0.001 | 0.834 | 18.387 | 0.000 |
| 2 | (Constant) | -4.685 | 6.595 | | -0.710 | 0.479 |
| | $X_{25}$ | 0.006 | 0.001 | 0.460 | 4.586 | 0.000 |
| | $X_5$ | 47.620 | 11.513 | 0.415 | 4.136 | 0.000 |
| 3 | (Constant) | -15.659 | 7.444 | | -2.104 | 0.037 |
| | $X_{25}$ | 0.006 | 0.001 | 0.442 | 4.516 | 0.000 |
| | $X_5$ | 44.958 | 11.265 | 0.391 | 3.991 | 0.000 |
| | $X_{27}$ | 3.923 | 1.340 | 0.129 | 2.928 | 0.004 |
| 4 | (Constant) | 57.838 | 26.267 | | 2.202 | 0.029 |
| | $X_{25}$ | 0.006 | 0.001 | 0.439 | 4.596 | 0.000 |
| | $X_5$ | 43.695 | 10.995 | 0.380 | 3.974 | 0.000 |
| | $X_{27}$ | 4.070 | 1.308 | 0.134 | 3.112 | 0.002 |
| | $X_{24}$ | -0.215 | 0.074 | -0.120 | -2.912 | 0.004 |
| 5 | (Constant) | 19.837 | 29.578 | | 0.671 | 0.504 |
| | $X_{25}$ | 0.005 | 0.001 | 0.393 | 4.127 | 0.000 |
| | $X_5$ | 41.672 | 10.809 | 0.363 | 3.855 | 0.000 |
| | $X_{27}$ | 3.981 | 1.283 | 0.131 | 3.103 | 0.002 |
| | $X_{24}$ | -0.223 | 0.072 | -0.125 | -3.079 | 0.002 |
| | $X_{14}$ | 10.966 | 4.197 | 0.122 | 2.613 | 0.010 |

a. Dependent Variable: $Y_1$

**Table No 6.6. Model Summary of stepwise regression of Socialization of pulse enterprise ($Y_1$) with 27 causal variables**

| Model | R | R Square | Adjusted R Square | Std. Error of the Estimate | Change Statistics | | | | |
|---|---|---|---|---|---|---|---|---|---|
| | | | | | R Square Change | F Change | df1 | df2 | Sig. F Change |
| 1 | 0.834[a] | 0.696 | 0.693 | 47.521016 | 0.696 | 338.068 | 1 | 148 | 0.000 |
| 2 | 0.853[b] | 0.727 | 0.724 | 45.128724 | 0.032 | 17.107 | 1 | 147 | 0.000 |
| 3 | 0.862[c] | 0.742 | 0.737 | 44.009339 | 0.015 | 8.573 | 1 | 146 | 0.004 |
| 4 | 0.870[d] | 0.757 | 0.750 | 42.923957 | 0.014 | 8.477 | 1 | 145 | 0.004 |
| 5 | 0.876[e] | 0.768 | 0.760 | 42.086691 | 0.011 | 6.827 | 1 | 144 | 0.010 |

## Results

Table No 6.5 and 6.6 Presents the step down multiple regression. It has been found that the variables gross return (Rs/ Bigha) ($X_{25}$), area under pulse cultivation (Bigha) ($X_5$), training Received ($X_{27}$), productivity or yield (Kg/Bigha) ($X_{24}$), farmer's attitude towards pulse crop cultivation ($X_{14}$) have been retained after eliminating the trivial in the preceding step.

Table No 6.6 presents these variables together have explained 76.8 per cent of total 'R' values of 78.7 per cent. So 5 variables of the total 27 variables merit highest importance in socialization of pulse enterprise ($Y_1$).

## Revelation

The variables retained in the last stage in stepwise regression analysis does present an operational constellation of 4 dominant variables working together and interacting reticulately it can characterize both the level and direction of disillusionment.

So, those few variables can go immensely important in making the farmers relinquished of disillusionment and thus have incubated an important strategic implementation for research locale and similes.

**Model No 6.2 Model of stepwise regression analysis of Socialization of pulse enterprise ($Y_1$) with 27 causal variables**

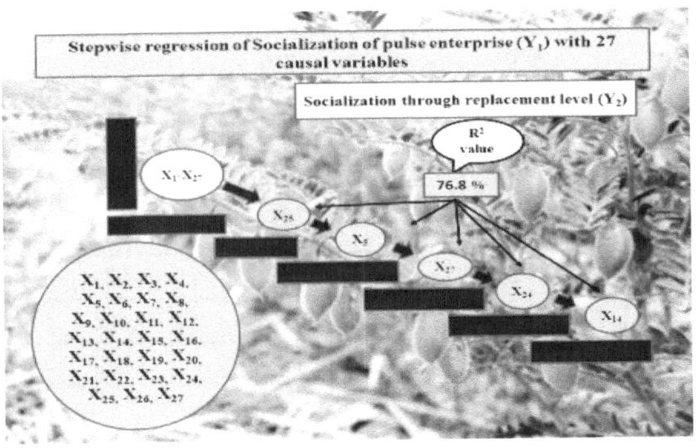

**Table No 6.7. Path analysis of comprehensive Socialization of pulse enterprise ($Y_1$) with 27 causal variables**

| Variables | Total effect | Direct effect | Indirect effect | Substantial effect | | |
|---|---|---|---|---|---|---|
| | | | | I | II | III |
| 1. Age ($X_1$) | -0.0433 | - 0.012 | -0.0310 | 0.0238 X5 | -0.0133 X14 | 0.0098 X24 |
| 2. Education ($X_2$) | 0.0312 | 0.0178 | 0.0133 | 0.0149 X25 | -0.0116 X15 | -0.0105 X5 |
| 3. Family size ($X_3$) | 0.2380 | -0.0433 | 0.2813 | 0.1299 X5 | 0.1053 X25 | 0.0144 X27 |
| 4. Family education (X4) | -0.0231 | 0.0138 | -0.0369 | -0.0159 X5 | -0.0145 X25 | -0.0125 X15 |
| 5. Area under pulse cultivation (Bigha) ($X_5$) | 0.8300 | 0.3609 (II) | 0.4691 (I) | 0.3549 X25 | 0.0564 X14 | 0.0352 X27 |
| 6. Farm size and technology adoption (Bigha) ($X_6$) | 0.0488 | -0.0141 | 0.0629 | 0.0381 X7 | 0.0224 X5 | -0.0083 X24 |
| 7. No of crop diversity ($X_7$) | -0.0935 | -0.0553 | -0.0382 | -0.0318 X5 | -0.0236 X25 | -0.0130 X15 |
| 8. Income (Rs /per capita/annum) ($X_8$) | -0.0415 | 0.0162 | -0.0578 | -0.0336 X24 | -0.0130 X25 | -0.0077 X3 |
| 9. Risk Orientation ($X_9$) | -0.0642 | 0.0272 | -0.0914 | -0.0444 X25 | -0.0437 X5 | -0.0149 X15 |
| 10. Scientific Orientation ($X_{10}$) | 0.0351 | -0.0629 | 0.0980 | 0.0285 X5 | 0.0228 X25 | 0.0223 X15 |
| 11. Planning orientation ($X_{11}$) | 0.0159 | 0.0319 | -0.0160 | -0.0198 X24 | 0.0152 X27 | -0.0077 X10 |
| 12. Production orientation ($X_{12}$) | -0.0141 | -0.0157 | 0.0016 | -0.0152 X5 | -0.0097 X15 | 0.0070 X24 |
| 13. Market Orientation ($X_{13}$) | 0.0901 | 0.0544 | 0.0357 | 0.0206 X5 | 0.0164 X27 | 0.0141 X25 |
| 14. Farmers attitude towards Pulse crop cultivation ($X_{14}$) | 0.5230 | 0.1163 | 0.4067 (III) | 0.1992 X25 | 0.1750 X5 | 0.0251 X15 |
| 15. Knowledge level of farmer towards cultivation of pulse crop ($X_{15}$) | 0.3910 | 0.0998 | 0.2912 | 0.1234 X25 | 0.1213 X5 | 0.0293 X14 |
| 16. Knowledge about insecticide ($X_{16}$) | -0.0078 | -0.0059 | -0.0018 | -0.0086 X25 | -0.0076 X15 | 0.0057 X24 |
| 17. Knowledge about fungicide ($X_{17}$) | 0.0657 | -0.0414 | 0.1071 | 0.0362 X25 | 0.0260 X5 | 0.0150 X15 |
| 18. Knowledge about weed control ($X_{18}$) | 0.0147 | 0.0232 | -0.0085 | -0.0125 X24 | -0.0097 X15 | -0.0071 X25 |
| 19. Knowledge about IPM practice ($X_{19}$) | 0.0504 | -0.0211 | 0.0716 | 0.0343 X5 | 0.0338 X25 | -0.0138 X24 |

| Variables | Total effect | Direct effect | Indirect effect | Substantial effect | | |
|---|---|---|---|---|---|---|
| | | | | I | II | III |
| 20. Farmers attitude towards IPM programme ($X_{20}$) | 0.0560 | 0.0037 | 0.0523 | 0.0275 X25 | 0.0245 X5 | -0.0105 X14 |
| 21. Attitude towards adoption ($X_{21}$) | 0.0486 | 0.0066 | 0.0420 | 0.0181 X24 | 0.0158 X15 | -0.0107 X13 |
| 22. Attitude towards discontinuous ($X_{22}$) | 0.0839 | -0.0205 | 0.1044 | 0.0365 X25 | 0.0260 X5 | 0.0130 X14 |
| 23. Attitude towards rejection ($X_{23}$) | -0.0276 | 0.0221 | -0.0497 | -0.0296 X5 | -0.0130 X25 | 0.0084 X10 |
| 24. Productivity or yield (kg/Bigha) ($X_{24}$) | -0.2020 | -0.1275 (III) | -0.0745 | -0.0385 X25 | -0.0375 X5 | -0.0170 X15 |
| 25. Gross return (Rs/ Bigha) ($X_{25}$) | 0.8340 | 0.3930 (I) | 0.4410 (II) | 0.3259 X5 | 0.0589 X14 | 0.0348 X27 |
| 26. Utilization/Information/ Cosmopolite ($X_{26}$) | -0.0653 | 0.0053 | -0.0706 | -0.0223 X15 | -0.0199 X5 | -0.0070 X13 |
| 27. Training received ($X_{27}$) | 0.3780 | 0.1170 | 0.2610 | 0.1167 X25 | 0.1086 X5 | 0.0208 X14 |

## Results

Table no 6.7 Presents the path analysis, by decomposing the total effect (r) of antecedent variables into direct indirect effect and residual effect. Path analysis has been administered to get direction and network of influence of antecedent variables on consequent variable.

From the table 6.7, it's is clear that variable, gross return (Rs/ Bigha) ($X_{25}$) has exerted highest direct effect on **socialization of pulse enterprise ($Y_1$)** followed by area under pulse cultivation (Bigha) ($X_5$) and productivity or yield (kg/Bigha) ($X_{24}$).

In case of indirect effect on **socialization of pulse enterprise ($Y_1$)** have area under pulse cultivation ($X_{25}$) and followed by gross return (Rs/ Bigha) ($X_{25}$) and farmer's attitude towards pulse crop cultivation ($X_{14}$).

It is discernible from the table the highest number of variables (21) has routed their substantial indirect effect through the variable, area under pulse cultivation (Bigha) ($X_5$). So, it could be inferred that these variables have got both substantive and associating properties to characterize the **socialization of pulse enterprise ($Y_1$)**.

Land resource as endowment is still the most important determinant for socialization for agriculture technology and the number of pulse crop. So better responses have been generated by those having higher land size while these didactic relation needs a support from attitude input from responds towards socialization of pulse enterprise ($Y_1$).

The residual effect being 0.2124, it is to conclude that 21.24 per cent of variation in this interaction could not be explained.

**Table No 6.8. Correlation coefficient of Socialization through replacement level ($Y_2$) with 27 independent variables**

| Variables | r- value | Sig. (2-tailed) |
|---|---|---|
| 1. Age ($X_1$) | -0.084 | 0.306 |
| 2. Education ($X_2$) | -0.052 | 0.527 |
| 3. Family size ($X_3$) | 0.120 | 0.144 |
| 4. Family education (X4) | 0.024 | 0.768 |
| 5. Area under pulse cultivation (Bigha) ($X_5$) | 0.458** | 0.000 |
| 6. Farm size and technology adoption (Bigha) ($X_6$) | -0.591** | 0.000 |
| 7. No of crop diversity ($X_7$) | 0.463** | 0.000 |
| 8. Income (Rs /per capita/annum) ($X_8$) | 0.064 | 0.438 |
| 9. Risk Orientation ($X_9$) | -0.161* | 0.049 |
| 10. Scientific Orientation ($X_{10}$) | 0.078 | 0.341 |
| 11. Planning orientation ($X_{11}$) | 0.082 | 0.317 |
| 12. Production orientation ($X_{12}$) | 0.103 | 0.210 |
| 13. Market Orientation ($X_{13}$) | 0.035 | 0.675 |
| 14. Farmers attitude towards Pulse crop cultivation ($X_{14}$) | 0.301** | 0.000 |
| 15. Knowledge level of farmer towards cultivation of pulse crop ($X_{15}$) | 0.105 | 0.202 |
| 16. Knowledge about insecticides ($X_{16}$) | 0.089 | 0.279 |
| 17. Knowledge about fungicide ($X_{17}$) | 0.155 | 0.059 |
| 18. Knowledge about weed control ($X_{18}$) | -0.043 | 0.604 |
| 19. Knowledge about IPM practice ($X_{19}$) | 0.130 | 0.112 |
| 20. Farmers attitude towards IPM programme ($X_{20}$) | -0.080 | 0.330 |
| 21. Attitude towards adoption ($X_{21}$) | 0.054 | 0.512 |
| 22. Attitude towards discontinuous ($X_{22}$) | 0.105 | 0.200 |
| 23. Attitude towards rejection ($X_{23}$) | -0.064 | 0.434 |
| 24. Productivity or yield (kg/Bigha) ($X_{24}$) | -0.115 | 0.160 |
| 25. Gross return (Rs/ Bigha) ($X_{25}$) | 0.471** | 0.000 |
| 26. Utilization of cosmopolite sources of information($X_{26}$) | 0.095 | 0.247 |
| 27. Training received ($X_{27}$) | 0.148 | 0.072 |

*Correlation is significant at 5 % level, **Correlation is significant at 1 % level.

## Results

Table No 6.8 presents the correlation coefficient of consequent variable, socialization of replacement level ($Y_2$) with 27 independent variables.

It has been found that variables viz. area under pulse cultivation (bigha) ($X_5$), no of crop diversity ($X_7$), farmer's attitude towards pulse crop cultivation ($X_{14}$), gross return (rs/ bigha) ($X_{25}$), have recorded the positive and farm size and technology adoption (bigha) ($X_6$), risk orientation ($X_9$) have recorded significant but negative correlation with socialization of replacement level ($Y_2$).

## Revelation

Replacement of land with pulse has been possible where there is more hand at disposal of farmer's. So, to bring more land on pulse crop depends on two things size of land under pulse and total land wide cultivation. It has also reflected higher cropping intensity but low risk orientation seems pulse in apparently easy growing crop, farmer with low risk orientation can also go safely, with pulse enterprises.

However, a positive attitude, as elicited by farmer's has helped positively growth pulse enterprises.

**Model No 6.3. Model of Correlation coefficient of Socialization through replacement level ($Y_2$) with 27 independent variables**

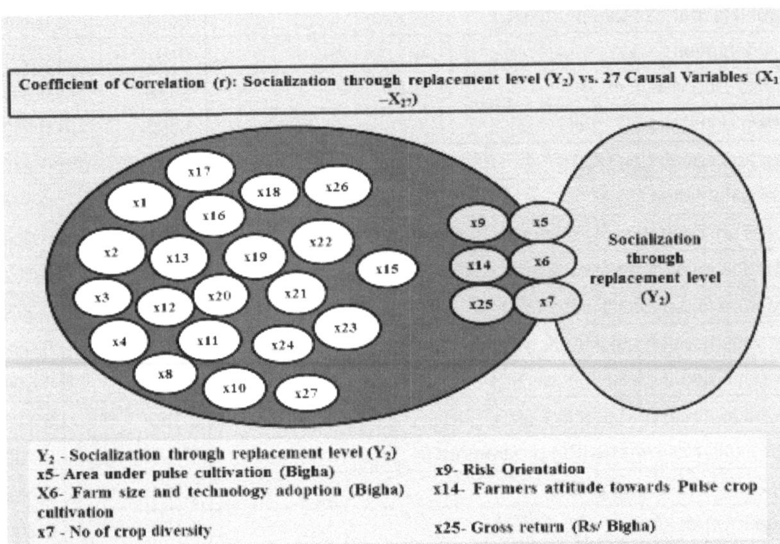


**Table No 6.9. Coefficients of Multiple regression analysis of Socialization through replacement level ($Y_2$) with 27 causal variables**

| Variables | Unstandardized Coefficients | | Standardized Coefficients | T | Sig. |
|---|---|---|---|---|---|
| | B | Std. Error | Beta | | |
| 1. Age ($X_1$) | -0.012 | 0.098 | -0.006 | -0.119 | 0.906 |
| 2. Education ($X_2$) | -0.291 | 0.414 | -0.040 | -0.702 | 0.484 |
| 3. Family size ($X_3$) | -0.244 | 0.245 | -0.059 | -0.996 | 0.321 |
| 4. Family education ($X_4$) | 0.683 | 0.489 | 0.080 | 1.397 | 0.165 |
| 5. Area under pulse cultivation (Bigha) ($X_5$) | 9.379 | 3.104 | 0.408 | 3.022 | 0.003 |
| 6. Farm size and technology adoption (Bigha) ($X_6$) | -1.652 | 0.263 | -0.483 | -6.294 | 0.000 |
| 7. No of crop diversity ($X_7$) | 5.798 | 2.676 | 0.163 | 2.167 | 0.032 |

| Variables | Unstandardized Coefficients | | Standardized Coefficients | T | Sig. |
|---|---|---|---|---|---|
| | B | Std. Error | Beta | | |
| 8. Income (Rs /per capita/annum) ($X_8$) | 7.652-005 | 0.000 | 0.092 | 1.646 | 0.102 |
| 9. Risk Orientation ($X_9$) | -1.404 | 0.806 | -0.096 | -1.741 | 0.084 |
| 10. Scientific Orientation ($X_{10}$) | -0.158 | 0.825 | -0.011 | -0.192 | 0.848 |
| 11. Planning orientation ($X_{11}$) | 0.005 | 0.819 | 0.000 | 0.007 | 0.995 |
| 12. Production orientation ($X_{12}$) | 1.053 | 0.775 | 0.073 | 1.358 | 0.177 |
| 13. Market Orientation ($X_{13}$) | 1.330 | 1.018 | 0.073 | 1.307 | 0.194 |
| 14. Farmers attitude towards Pulse crop cultivation ($X_{14}$) | 0.780 | 1.122 | 0.044 | 0.696 | 0.488 |
| 15. Knowledge level of farmer towards cultivation of pulse crop ($X_{15}$) | -0.387 | 0.493 | -0.050 | -0.785 | 0.434 |
| 16. Knowledge about insecticides ($X_{16}$) | 2.287 | 4.467 | 0.028 | 0.512 | 0.610 |
| 17. Knowledge about fungicide ($X_{17}$) | 8.515 | 4.595 | 0.102 | 1.853 | 0.066 |
| 18. Knowledge about weed control ($X_{18}$) | 0.445 | 2.863 | 0.009 | 0.156 | 0.877 |
| 19. Knowledge about IPM practice ($X_{19}$) | 15.264 | 6.751 | 0.121 | 2.261 | 0.026 |
| 20. Farmers attitude towards IPM programme ($X_{20}$) | -1.613 | 1.259 | -0.070 | -1.281 | 0.203 |
| 21. Attitude towards adoption ($X_{21}$) | 0.694 | 1.120 | 0.035 | 0.619 | 0.537 |
| 22. Attitude towards discontinuous ($X_{22}$) | 1.739 | 0.964 | 0.102 | 1.804 | 0.074 |
| 23. Attitude towards rejection ($X_{23}$) | 0.609 | 1.805 | 0.018 | 0.337 | 0.736 |
| 24. Productivity or yield (kg/Bigha) ($X_{24}$) | -0.027 | 0.021 | -0.077 | -1.320 | 0.189 |
| 25. Gross return (Rs/ Bigha) ($X_{25}$) | 0.000 | 0.000 | 0.108 | 0.829 | 0.409 |
| 26. Utilization of cosmopolite sources of information($X_{26}$) | 4.897 | 4.310 | 0.063 | 1.136 | 0.258 |
| 27. Training received ($X_{27}$) | -0.444 | 0.338 | -0.073 | -1.315 | 0.191 |

**Table No 6.10. Multiple regression analysis of Socialization through replacement level ($Y_2$) with 27 causal variables (Model Summary**

| Model | R | R Square | Adjusted R Square | Std. Error of the Estimate | Change Statistics | | | | |
|---|---|---|---|---|---|---|---|---|---|
| | | | | | R Square Change | F Change | df1 | df2 | Sig. F Change |
| 1 | 0.830[a] | 0.689 | 0.620 | 10.573862 | 0.689 | 10.019 | 27 | 122 | 0.000 |

**Results**

Table No 6.9. Presents the multiple regression analysis which reflects the functional efficacy of the correlation through 'beta' value and respective 't' values of the casual variables on the consequent variables i.e. socialization through replacement level ($Y_2$)

Table No 6.10. The $R^2$ value 0.689 being, it's is to conclude that even combination of all 27 variables so far, 68.9 % of variability embedded with the consequent variable has been explained.

**Table No 6.11. Stepwise regression analysis of Socialization through replacement level ($Y_2$) with 27 causal variables**

| Model | | Unstandardized Coefficients | | Standardized Coefficients | t | Sig. |
|---|---|---|---|---|---|---|
| | | B | Std. Error | Beta | | |
| 1 | (Constant) | 34.815 | 2.343 | | 14.862 | 0.000 |
| | $X_6$ | -2.023 | 0.227 | -0.591 | -8.909 | 0.000 |
| 2 | (Constant) | 24.491 | 2.155 | | 11.363 | 0.000 |
| | $X_6$ | -2.128 | 0.180 | -0.622 | -11.807 | 0.000 |
| | $X_5$ | 11.394 | 1.209 | 0.496 | 9.422 | 0.000 |
| 3 | (Constant) | 8.175 | 7.119 | | 1.148 | 0.253 |
| | $X_6$ | -2.163 | 0.178 | -0.632 | -12.154 | 0.000 |
| | $X_5$ | 11.203 | 1.193 | 0.488 | 9.393 | 0.000 |
| | $X_{22}$ | 2.135 | 0.889 | 0.125 | 2.401 | 0.018 |
| 4 | (Constant) | -12.992 | 12.271 | | -1.059 | 0.292 |
| | $X_6$ | -2.173 | 0.176 | -0.635 | -12.345 | 0.000 |
| | $X_5$ | 10.959 | 1.185 | 0.477 | 9.251 | 0.000 |
| | $X_{22}$ | 2.252 | 0.880 | 0.132 | 2.557 | 0.012 |
| | $X_{19}$ | 13.660 | 6.488 | 0.108 | 2.105 | 0.037 |

a. Dependent Variable: $Y_2$

**Table No 6.12. Model Summary of stepwise regression of Socialization through replacement level ($Y_2$) with 27 causal variables**

| Model | R | R Square | Adjusted R Square | Std. Error of the Estimate | Change Statistics | | | | |
|---|---|---|---|---|---|---|---|---|---|
| | | | | | R Square Change | F Change | df1 | df2 | Sig. F Change |
| 1 | 0.591[a] | 0.349 | 0.345 | 13.893183 | 0.349 | 79.368 | 1 | 148 | 0.000 |
| 2 | 0.771[b] | 0.594 | 0.589 | 11.007227 | 0.245 | 88.781 | 1 | 147 | 0.000 |
| 3 | 0.781[c] | 0.610 | 0.602 | 10.833025 | 0.015 | 5.766 | 1 | 146 | 0.018 |
| 4 | 0.788[d] | 0.621 | 0.611 | 10.707889 | 0.012 | 4.432 | 1 | 145 | 0.037 |

## Results

Table No 6.11 and 6.12 Presents the step down multiple regression. It has been found that the variables farm size and technology adoption (bigha) ($X_6$), area under pulse cultivation (bigha) ($X_5$), attitude towards discontinuous ($X_{22}$), knowledge about IPM practice ($X_{19}$) have been retained after eliminating the trivial in the preceding step.

So these variables together have explained 62.1 per cent of total '$r^2$' values of 68.9 per cent. So 4 variables of the total 27 variables merit highest importance in socialization through replacement level ($Y_2$).

## Revelation

The variables retained in the last stage in stepwise regression analysis does present and operational constellation of 4 dominant variables working together and interacting ridiculously it can characterize both the level and direction of disillusionment.

So, those few variables can go immensely important in making the farmers relinquished of disillusionment and thus have incubated an important strategic implementation for research locale and similes.

**Model No 6.4. Model of Stepwise regression analysis of Socialization through replacement level ($Y_2$) with 27 causal variables**

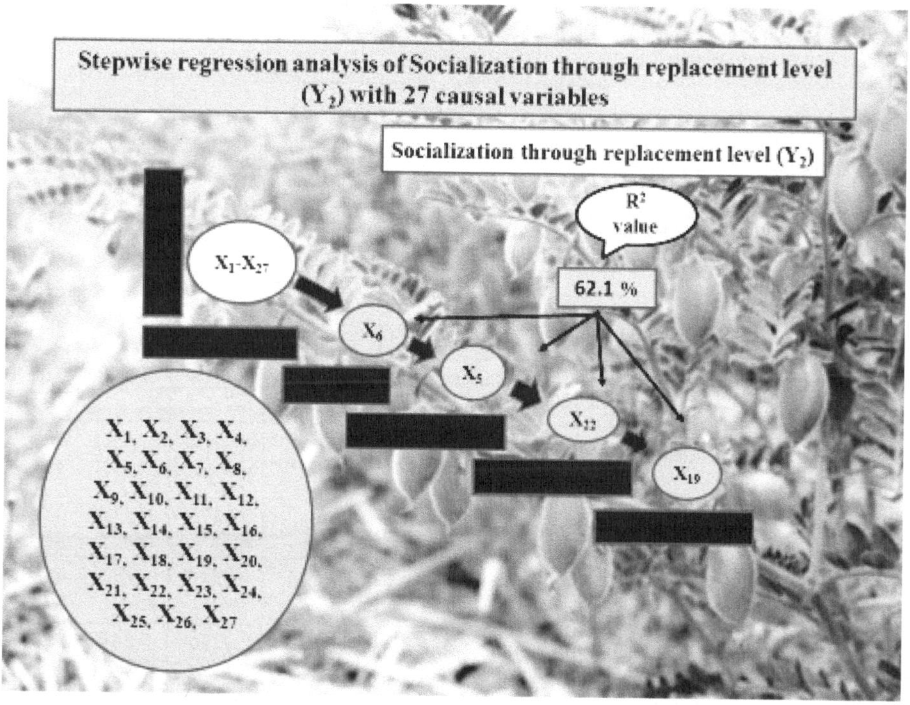

**Table No 6.13. Path analysis of comprehensive Socialization through replacement level ($Y_2$) with 27 causal variables**

| Variables | Total effect | Direct effect | Indirect effect | Substantial effect | | |
|---|---|---|---|---|---|---|
| | | | | I | II | III |
| 1. Age ($X_1$) | -0.0841 | -0.0060 | -0.0782 | -0.0273 X5 | -0.0101 X6 | -0.0089 X22 |
| 2. Education ($X_2$) | -0.0520 | -0.0398 | -0.0122 | -0.0410 X6 | 0.0278 X4 | -0.0120 X5 |
| 3. Family size ($X_3$) | 0.1199 | -0.0596 | 0.1795 | 0.1487 X5 | 0.0278 X25 | -0.0227 X6 |
| 4. Family education (X4) | 0.0243 | 0.0806 | -0.0564 | -0.0362 X6 | 0.0209 X22 | -0.0182 X5 |
| 5. Area under pulse cultivation (Bigha) ($X_5$) | 0.4580 | 0.4130 (II) | 0.0450 | 0.0938 X25 | -0.0299 X6 | -0.0220 X27 |
| 6. Farm size and technology adoption (Bigha) ($X_6$) | -0.5910 | -0.4829 (I) | -0.1081 | -0.1125 X7 | 0.0256 X5 | 0.0087 X22 |
| 7. Number of crop diversity ($X_7$) | 0.4630 | 0.1635 | 0.2995(II) | 0.1635 X7 | -0.0363 X5 | -0.0088 X17 |
| 8. Income (Rs /per capita/annum) ($X_8$) | 0.0638 | 0.0921 | -0.0283 | -0.0201 X24 | -0.0105 X3 | 0.0077 X7 |
| 9. Risk Orientation ($X_9$) | -0.1610 | -0.0966 | -0.0644 | -0.0500 X5 | -0.0353 X6 | -0.0117 X25 |
| 10. Scientific Orientation ($X_{10}$) | 0.0783 | -0.0117 | 0.0900 | 0.0483 X6 | 0.0326 X5 | -0.0165 X22 |
| 11. Planning orientation ($X_{11}$) | 0.0823 | 0.0005 | 0.0818 | 0.0642 X6 | 0.0190 X7 | -0.0118 X24 |
| 12. Production orientation ($X_{12}$) | 0.1030 | 0.0728 | 0.0301 | 0.0449 X6 | -0.0173 X5 | -0.0090 X19 |
| 13. Market Orientation ($X_{13}$) | 0.0345 | 0.0731 | -0.0386 | -0.0425 X6 | 0.0235 X5 | 0.0143 X4 |
| 14. Farmers attitude towards Pulse crop cultivation ($X_{14}$) | 0.3010 | 0.0436 | 0.2574 (III) | 0.2003 X5 | 0.0526 X25 | -0.0131 X27 |
| 15. Knowledge level of farmer towards cultivation of pulse crop ($X_{15}$) | 0.1048 | -0.0492 | 0.1540 | 0.1388 X5 | 0.0326 X25 | -0.0212 X7 |
| 16. Knowledge about insecticide ($X_{16}$) | 0.0890 | 0.0273 | 0.0617 | 0.0372 X6 | 0.0144 X19 | 0.0132 X7 |
| 17. Knowledge about fungicide ($X_{17}$) | 0.1547 | 0.1029 | 0.0518 | 0.0314 X6 | -0.0141 X7 | 0.0096 X25 |
| 18. Knowledge about weed control ($X_{18}$) | -0.0427 | 0.0086 | -0.0514 | -0.0328 X6 | -0.0234 X9 | 0.0158 X17 |
| 19. Knowledge about IPM practice ($X_{19}$) | 0.1303 | 0.1210 (III) | 0.0093 | 0.0392 X5 | -0.0126 X6 | 0.0089 X25 |
| 20. Farmers attitude towards IPM programme ($X_{20}$) | -0.0801 | -0.0702 | -0.0099 | -0.0362 X6 | 0.0281 X5 | -0.0139 X7 |
| 21. Attitude towards adoption ($X_{21}$) | 0.0540 | 0.0343 | 0.0196 | 0.0266 X6 | -0.0143 X13 | -0.0125 X19 |

| Variables | Total effect | Direct effect | Indirect effect | Substantial effect | | |
|---|---|---|---|---|---|---|
| | | | | I | II | III |
| 22. Attitude towards discontinuous ($X_{22}$) | 0.1052 | 0.1013 | 0.0038 | -0.0415 X6 | 0.0297 X5 | 0.0166 X4 |
| 23. Attitude towards rejection ($X_{23}$) | -0.0643 | 0.0184 | -0.0827 | -0.0339 X5 | -0.0164 X6 | -0.0125 X22 |
| 24. Productivity or yield (kg/Bigha) ($X_{24}$) | -0.1153 | -0.0762 | -0.0390 | -0.0429 X5 | -0.0314 X6 | 0.0243 X8 |
| 25. Gross return (Rs/ Bigha) ($X_{25}$) | 0.4710 | 0.1038 | 0.3672 (I) | 0.3729 X5 | 0.0221 X14 | -0.0218 X27 |
| 26. Utilization/Information/ Cosmopolite ($X_{26}$) | 0.0950 | 0.0628 | 0.0323 | 0.0633 X6 | -0.0227 X5 | 0.0110 X15 |
| 27. Training received ($X_{27}$) | 0.1476 | -0.0732 | 0.2208 | 0.1243 X5 | 0.0308 X25 | 0.0270 X6 |

## Results

Table No 6.13 presents the path analysis, by decomposing the total effect (r) of antecedent variables into direct indirect effect and residual effect. Path analysis has been administered to get direction and network of influence of antecedent variables on consequent variable.

From the table no 6.13, it's is clear that variable, farm size and technology adoption (Bigha) ($X_6$) has exerted highest direct effect on **socialization through replacement level ($Y_2$)** followed by area under pulse cultivation (Bigha) ($X_5$) and knowledge about IPM practice ($X_{19}$).

In case of indirect effect on **socialization through replacement level ($Y_2$)** area under pulse cultivation (Bigha) ($X_5$) has exerted the highest indirect effect on **socialization through replacement level ($Y_2$)** followed by number of crop diversity ($X_7$) and farmer's attitude towards pulse crop cultivation ($X_{14}$).

It is discernible from the table the highest number of variables (20) has routed their substantial indirect effect through the variable, Farm size and technology adoption (Bigha) ($X_6$). So, it could be inferred that these variables have got both substantive and associational properties to characterize the **socialization through replacement level ($Y_2$).**

The residual effect being 0.3102, it is to conclude that 31.02 per cent of variation in this relation could not be explained.

**Table No 6.14. Correlation coefficient of Socialization through reinvention ($Y_3$) with 27 independent variables**

| Variables | r- value | Sig. (2-tailed) |
|---|---|---|
| 1. Age ($X_1$) | 0.085 | 0.302 |
| 2. Education ($X_2$) | 0.023 | 0.777 |
| 3. Family size ($X_3$) | 0.032 | 0.697 |

| Variables | r- value | Sig. (2-tailed) |
|---|---|---|
| 4. Family education (X4) | -0.002 | 0.982 |
| 5. Area under pulse cultivation (Bigha) ($X_5$) | 0.093 | 0.256 |
| 6. Farm size and technology adoption (Bigha) ($X_6$) | -0.156 | 0.056 |
| 7. No of crop diversity ($X_7$) | 0.140 | 0.088 |
| 8. Income (Rs /per capita/annum) ($X_8$) | 0.038 | 0.644 |
| 9. Risk Orientation ($X_9$) | -0.005 | 0.947 |
| 10. Scientific Orientation ($X_{10}$) | 0.043 | 0.599 |
| 11. Planning orientation ($X_{11}$) | 0.147 | 0.072 |
| 12. Production orientation ($X_{12}$) | -0.027 | 0.745 |
| 13. Market Orientation ($X_{13}$) | -0.006 | 0.940 |
| 14. Farmers attitude towards Pulse crop cultivation ($X_{14}$) | 0.081 | 0.324 |
| 15. Knowledge level of farmer towards cultivation of pulse crop ($X_{15}$) | 0.099 | 0.226 |
| 16. Knowledge about insecticides ($X_{16}$) | 0.076 | 0.354 |
| 17. Knowledge about fungicide ($X_{17}$) | 0.019 | 0.817 |
| 18. Knowledge about weed control ($X_{18}$) | 0.125 | 0.127 |
| 19. Knowledge about IPM practice ($X_{19}$) | 0.063 | 0.445 |
| 20. Farmers attitude towards IPM programme ($X_{20}$) | -0.039 | 0.640 |
| 21. Attitude towards adoption ($X_{21}$) | .195* | 0.017 |
| 22. Attitude towards discontinuous ($X_{22}$) | -0.060 | 0.465 |
| 23. Attitude towards rejection ($X_{23}$) | 0.081 | 0.325 |
| 24. Productivity or yield (kg/Bigha) ($X_{24}$) | -0.019 | 0.814 |
| 25. Gross return (Rs/ Bigha) ($X_{25}$) | 0.138 | 0.093 |
| 26. Utilization of cosmopolite sources of information($X_{26}$) | 0.122 | 0.138 |
| 27. Training received ($X_{27}$) | .509** | 0.000 |

*Correlation is significant at 5 % level, **Correlation is significant at 1 % level.

## Results

Table No 6.14. presents the correlation coefficient of consequent variable, socialization of reinvention ($Y_3$) with 27 independent variables.

It has been found that variables viz. attitude towards adoption ($X_{21}$), training received ($X_{27}$) have recorded positive correlation with socialization through reinvention ($Y_3$).

## Revelation

For making the reinvention a real execution, attitude towards adoption of individual is very much needed. Adoption provide an opportunity to realize an innovation to happen. An adoption of any new idea or new innovation leads to complete socialization of reinvention, and here, it has shown for the socialization of reinvention.

Training help in enabling an individual technically sound along with knowledge and skill about particular modification of an innovation. That why, all these two variables have gone positively and significantly correlated with the socialization through reinvention.

**Model No 6.5. Model of Correlation coefficient of Socialization through reinvention ($Y_3$) with 27 independent variables**

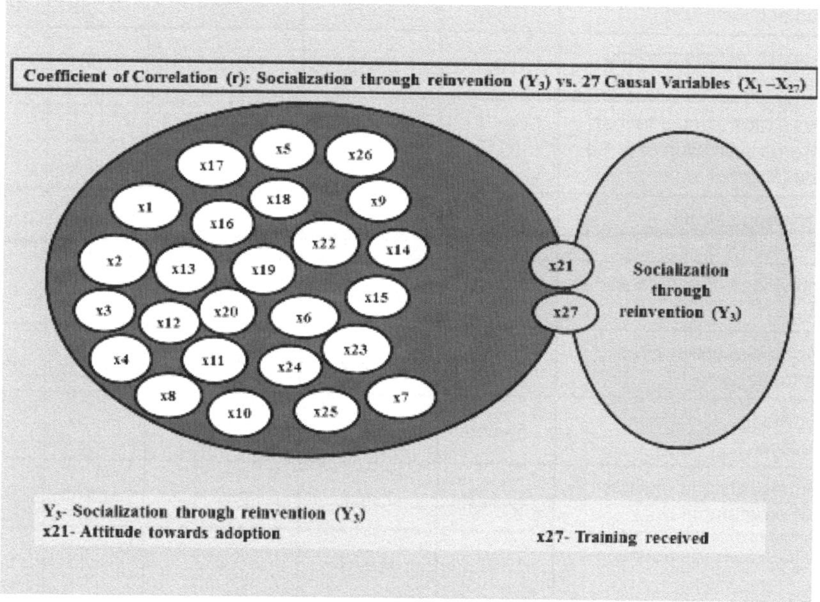


**Table No 6.15. Coefficients of Multiple regression analysis of Socialization through reinvention ($Y_3$) with 27 causal variables**

| Variables | Unstandardized Coefficients | | Standardized Coefficients | T | Sig. |
|---|---|---|---|---|---|
| | B | Std. Error | Beta | | |
| 1. Age ($X_1$) | 0.534 | 0.605 | 0.068 | 0.883 | 0.379 |
| 2. Education ($X_2$) | -0.539 | 2.560 | -0.017 | -0.211 | 0.834 |
| 3. Family size ($X_3$) | -0.269 | 1.514 | -0.015 | -0.178 | 0.859 |
| 4. Family education ($X_4$) | 1.378 | 3.022 | 0.037 | 0.456 | 0.649 |
| 5. Area under pulse cultivation (Bigha) ($X_5$) | -27.915 | 19.180 | -0.278 | -1.455 | 0.148 |
| 6. Farm size and technology adoption (Bigha) ($X_6$) | -1.022 | 1.622 | -0.068 | -0.630 | 0.530 |
| 7. No of crop diversity ($X_7$) | 3.596 | 16.536 | 0.023 | 0.217 | 0.828 |
| 8. Income (Rs /per capita/ annum) ($X_8$) | 0.000 | 0.000 | 0.067 | 0.845 | 0.400 |
| 9. Risk Orientation ($X_9$) | -1.386 | 4.983 | -0.022 | -0.278 | 0.781 |

| Variables | Unstandardized Coefficients | | Standardized Coefficients | T | Sig. |
|---|---|---|---|---|---|
| | B | Std. Error | Beta | | |
| 10. Scientific Orientation ($X_{10}$) | 0.284 | 5.098 | 0.005 | 0.056 | 0.956 |
| 11. Planning orientation ($X_{11}$) | 3.496 | 5.063 | 0.054 | 0.691 | 0.491 |
| 12. Production orientation ($X_{12}$) | -2.823 | 4.791 | -0.045 | -0.589 | 0.557 |
| 13. Market Orientation ($X_{13}$) | -1.268 | 6.289 | -0.016 | -0.202 | 0.841 |
| 14. Farmers attitude towards Pulse crop cultivation ($X_{14}$) | -0.424 | 6.930 | -0.005 | -0.061 | 0.951 |
| 15. Knowledge level of farmer towards cultivation of pulse crop ($X_{15}$) | 1.115 | 3.047 | 0.033 | 0.366 | 0.715 |
| 16. Knowledge about insecticides ($X_{16}$) | 11.618 | 27.606 | 0.032 | 0.421 | 0.675 |
| 17. Knowledge about fungicide ($X_{17}$) | -24.600 | 28.397 | -0.068 | -0.866 | 0.388 |
| 18. Knowledge about weed control ($X_{18}$) | 34.136 | 17.692 | 0.158 | 1.930 | 0.056 |
| 19. Knowledge about IPM practice ($X_{19}$) | 55.568 | 41.717 | 0.101 | 1.332 | 0.185 |
| 20. Farmers attitude towards IPM programme ($X_{20}$) | -3.457 | 7.781 | -0.034 | -0.444 | 0.658 |
| 21. Attitude towards adoption ($X_{21}$) | 13.448 | 6.923 | 0.153 | 1.943 | 0.054 |
| 22. Attitude towards discontinuous ($X_{22}$) | -6.644 | 5.957 | -0.089 | -1.115 | 0.267 |
| 23. Attitude towards rejection ($X_{23}$) | 2.326 | 11.154 | 0.016 | 0.209 | 0.835 |
| 24. Productivity or yield (kg/ Bigha) ($X_{24}$) | -0.095 | 0.128 | -0.060 | -0.738 | 0.462 |
| 25. Gross return (Rs/ Bigha) ($X_{25}$) | 0.003 | 0.002 | 0.249 | 1.347 | 0.181 |
| 26. Utilization of cosmopolite sources of information($X_{26}$) | 40.019 | 26.631 | 0.117 | 1.503 | 0.135 |
| 27. Training received ($X_{27}$) | 13.509 | 2.087 | 0.508 | 6.472 | 0.000 |

**Table No 6.16. Multiple regression analysis of Socialization through reinvention ($Y_3$) with 27 causal variables (Model Summary)**

| Model | R | R Square | Adjusted R Square | Std. Error of the Estimate | Change Statistics | | | | |
|---|---|---|---|---|---|---|---|---|---|
| | | | | | R Square Change | F Change | df1 | df2 | Sig. F Change |
| 1 | 0.617[a] | 0.381 | 0.244 | 65.339825 | 0.381 | 2.780 | 27 | 122 | 0.000 |

## Results

Table No 6.15. Presents the multiple regression analysis which reflects the functional efficacy of the correlation through 'beta' value and respective 't' values of the casual variables on the consequent variables i.e socialization through reinvention ($Y_3$)

Table No 6.16. The $R^2$ value 0.381 being, it's is to conclude that even combination of all 27 variables so far, 38.1 % of variability embedded with the consequent variable has been explained.

**Table No 6.17. Stepwise regression analysis of Socialization through reinvention ($Y_3$) with 27 causal variables**

| Model | | Unstandardized Coefficients | | Standardized Coefficients | t | Sig. |
|---|---|---|---|---|---|---|
| | | B | Std. Error | Beta | | |
| 1 | (Constant) | 71.684 | 9.255 | | 7.746 | 0.000 |
| | $X_{27}$ | 13.522 | 1.881 | 0.509 | 7.189 | 0.000 |
| 2 | (Constant) | -11.761 | 34.772 | | -0.338 | 0.736 |
| | $X_{27}$ | 13.325 | 1.851 | 0.501 | 7.200 | 0.000 |
| | $X_{21}$ | 15.215 | 6.119 | 0.173 | 2.486 | 0.014 |

a. Dependent Variable: $Y_3$

**Table No 6.18. Model Summary of stepwise regression of Socialization through reinvention ($Y_3$) with 27 causal variables**

| Model | R | R Square | Adjusted R Square | Std. Error of the Estimate | Change Statistics | | | | |
|---|---|---|---|---|---|---|---|---|---|
| | | | | | R Square Change | F Change | df1 | df2 | Sig. F Change |
| 1 | 0.509[a] | 0.259 | 0.254 | 64.910152 | 0.259 | 51.677 | 1 | 148 | 0.000 |
| 2 | 0.537[b] | 0.289 | 0.279 | 63.802814 | 0.030 | 6.182 | 1 | 147 | 0.014 |

## Results

Table No 6.17 presents the step down multiple regression. It has been found that the variables training received ($X_{27}$), attitude towards adoption ($X_{21}$) have been retained after eliminating the trivial in the preceding step.

Table No 6.18 presents So these variables together have explained 27.9 per cent of total '$R^2$' values of 38.1 per cent. So 2 variables of the total 27 variables merit highest importance in socialization through reinvention ($Y_3$).

## Revelation

The variables retained in the last stage in stepwise regression analysis does present and operational constellation of 2 dominant variables working together and interacting ridiculously it can characterize both the level and direction of disillusionment.

So, those few variables can go immensely important in making the farmers relinquished of disillusionment and thus have incubated an important strategic implementation for research locale and similes.

**Model No 6.6. Model of Stepwise regression analysis of Socialization through reinvention ($Y_3$) with 27 causal variables**

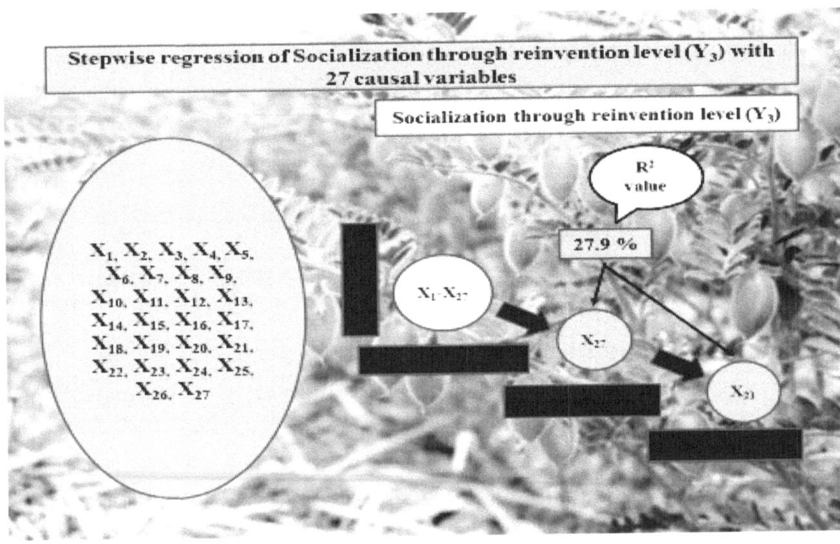


**Table No 6.19. Path analysis of comprehensive Socialization through reinvention level ($Y_3$) with 27 causal variables**

| Variables | Total effect | Direct effect | Indirect effect | Substantial effect | | |
|---|---|---|---|---|---|---|
| | | | | I | I | I |
| 1. Age ($X_1$) | 0.0848 | 0.0680 | 0.0168 | 0.0182 X5 | 0.0153 X27 | -0.0098 X26 |
| 2. Education ($X_2$) | 0.0233 | -0.0171 | 0.0404 | 0.0376 X27 | 0.0128 X4 | 0.0094 X25 |
| 3. Family size ($X_3$ | 0.0320 | -0.0152 | 0.0472 | -0.0993 X5 | 0.0662 X25 | 0.0625 X27 |
| 4. Family education (X4) | -0.0018 | 0.0372 | -0.0390 | -0.0182 X22 | -0.0167 X21 | 0.0137 X27 |
| 5. Area under pulse cultivation (Bigha) ($X_5$) | 0.0934 | -0.2758 (II) | 0.3692 (I) | 0.2230 X25 | 0.1531 X27 | 0.0110 X15 |
| 6. Farm size and technology adoption (Bigha) ($X_6$) | -0.1563 | -0.0684 | -0.0879 | -0.0285 X27 | -0.0171 X5 | -0.0159 X7 |

| Variables | Total effect | Direct effect | Indirect effect | Substantial effect | | |
|---|---|---|---|---|---|---|
| | | | | I | I | I |
| 7. Number of crop diversity ($X_7$) | 0.1400 | 0.0231 | 0.1169 (II) | 0.0471 X6 | 0.0386 X27 | 0.0243 X5 |
| 8. Income (Rs /per capita/annum) ($X_8$) | 0.0381 | 0.0004 | -0.0283 | -0.0158 X24 | -0.0116 X26 | -0.0002 X25 |
| 9. Risk Orientation ($X_9$) | -0.0054 | -0.0211 | 0.0156 | 0.0381 X18 | 0.0334 X5 | -0.0279 X25 |
| 10. Scientific Orientation ($X_{10}$) | 0.0433 | 0.0049 | 0.0384 | 0.0269 X27 | -0.0218 X5 | 0.0144 X22 |
| 11. Planning orientation ($X_{11}$) | 0.1473 | 0.0536 | 0.0937 | 0.0661 X27 | 0.0536 X5 | 0.0274 X18 |
| 12. Production orientation ($X_{12}$) | -0.0268 | -0.0444 | 0.0176 | 0.0147 X27 | 0.0116 X5 | -0.0084 X26 |
| 13. Market Orientation ($X_{13}$) | -0.0062 | -0.0164 | 0.0101 | 0.0712 X27 | -0.0301 X21 | -0.0157 X5 |
| 14. Farmers attitude towards Pulse crop cultivation ($X_{14}$) | 0.0811 | -0.0057 | 0.0868 | -0.1338 X5 | 0.1252 X25 | 0.0910 X27 |
| 15. Knowledge level of farmer towards cultivation of pulse crop ($X_{15}$) | 0.0994 | 0.0328 | 0.0666 | 0.0997 X27 | -0.0927 X5 | 0.0775 X25 |
| 16. Knowledge about insecticide ($X_{16}$) | 0.0761 | 0.0319 | 0.0443 | 0.0184 X26 | 0.0120 X19 | 0.0112 X27 |
| 17. Knowledge about fungicide ($X_{17}$) | 0.0190 | -0.0671 | 0.0862 | 0.0397 X27 | 0.0243 X18 | 0.0227 X25 |
| 18. Knowledge about weed control ($X_{18}$) | 0.1253 | 0.1576 | -0.0323 | 0.0105 X26 | -0.0103 X17 | 0.0093 X11 |
| 19. Knowledge about IPM practice ($X_{19}$) | 0.0629 | 0.1007 | -0.0379 | -0.0262 X5 | 0.0212 X25 | -0.0158 X21 |
| 20. Farmers attitude towards IPM programme ($X_{20}$) | -0.0385 | -0.0345 | -0.0041 | 0.0331 X18 | -0.0264 X27 | -0.0209 X21 |
| 21. Attitude towards adoption ($X_{21}$) | 0.1950 | 0.1534 | 0.0416 | 0.0219 X27 | -0.0104 X19 | 0.0085 X24 |
| 22. Attitude towards discontinuous ($X_{22}$) | -0.0601 | -0.0884 | 0.0283 | 0.0366 X27 | 0.0230 X25 | -0.0199 X5 |
| 23. Attitude towards rejection ($X_{23}$) | 0.0809 | 0.0163 | 0.0646 | 0.0226 X5 | 0.0219 X27 | -0.0078 X18 |
| 24. Productivity or yield (kg/Bigha) ($X_{24}$) | -0.0194 | -0.0597 | 0.0404 | 0.0287 X5 | -0.0242 X25 | -0.0218 X21 |
| 25. Gross return (Rs/ Bigha) ($X_{25}$) | 0.1376 | 0.2470 (III) | -0.1094 (III) | -0.2491 X5 | 0.1510 X27 | 0.0103 X15 |

| Variables | Total effect | Direct effect | Indirect effect | Substantial effect | | |
|---|---|---|---|---|---|---|
| | | | | I | I | I |
| 26. Utilization/ Information/ Cosmopolite ($X_{26}$) | 0.1218 | 0.1169 | 0.0049 | 0.0142 X18 | -0.0126 X25 | -0.0092 X27 |
| 27. Training received ($X_{27}$) | 0.5090 | 0.5085 (I) | 0.0005 | -0.0830 X5 | 0.0734 X25 | 0.0070 X11 |

## Results

Table No 6.19 presents the path analysis, by decomposing the total effect (r) of antecedent variables into direct indirect effect and residual effect. Path analysis has been administered to get direction and network of influence of antecedent variables on consequent variable.

From the table, it's is clear that variable, training received ($X_{27}$) has exerted highest direct effect on **socialization through reinvention level ($Y_3$)** followed by area under pulse cultivation (Bigha) ($X_5$) and gross return (Rs/ Bigha) ($X_{25}$).

In case of indirect effect on **socialization through reinvention level ($Y_3$)** area under pulse cultivation (Bigha) ($X_5$) has exerted the highest indirect effect on **socialization through reinvention level ($Y_3$)** followed by number of crop diversity ($X_7$) and gross return (Rs/ Bigha) ($X_{25}$).

It is discernible from the table the highest number of variables (22) has routed their substantial indirect effect through the variable, training received ($X_{27}$). So, it could be inferred that these variables have got both substantive and associational properties to characterize the **socialization through reinvention level ($Y_3$).**

The residual effect being 0.6188, it is to conclude that 61.88 per cent of variation in this relation could not be explained.

**Table No 6.20. Correlation coefficient of Marketed surplus ($Y_4$) with 27 independent variables**

| Variables | r- value | Sig. (2-tailed) |
|---|---|---|
| 1. Age ($X_1$) | 0.010 | 0.907 |
| 2. Education ($X_2$) | 0.052 | 0.531 |
| 3. Family size ($X_3$) | -0.054 | 0.513 |
| 4. Family education (X4) | 0.024 | 0.775 |
| 5. Area under pulse cultivation (Bigha) ($X_5$) | -0.092 | 0.265 |
| 6. Farm size and technology adoption (Bigha) ($X_6$) | -0.052 | 0.524 |
| 7. No of crop diversity ($X_7$) | 0.026 | 0.753 |
| 8. Income (Rs /per capita/annum) ($X_8$) | 0.215** | 0.008 |
| 9. Risk Orientation ($X_9$) | -0.004 | 0.964 |
| 10. Scientific Orientation ($X_{10}$) | -0.009 | 0.911 |

| Variables | r- value | Sig. (2-tailed) |
|---|---|---|
| 11. Planning orientation ($X_{11}$) | 0.132 | 0.107 |
| 12. Production orientation ($X_{12}$) | -0.006 | 0.943 |
| 13. Market Orientation ($X_{13}$) | 0.161* | 0.050 |
| 14. Farmers attitude towards Pulse crop cultivation ($X_{14}$) | -0.120 | 0.143 |
| 15. Knowledge level of farmer towards cultivation of pulse crop ($X_{15}$) | -0.106 | 0.198 |
| 16. Knowledge about insecticides ($X_{16}$) | 0.001 | 0.989 |
| 17. Knowledge about fungicide ($X_{17}$) | -0.017 | 0.834 |
| 18. Knowledge about weed control ($X_{18}$) | 0.140 | 0.088 |
| 19. Knowledge about IPM practice ($X_{19}$) | 0.019 | 0.817 |
| 20. Farmers attitude towards IPM programme ($X_{20}$) | -0.034 | 0.681 |
| 21. Attitude towards adoption ($X_{21}$) | -0.189* | 0.021 |
| 22. Attitude towards discontinuous ($X_{22}$) | 0.049 | 0.550 |
| 23. Attitude towards rejection ($X_{23}$) | -0.020 | 0.803 |
| 24. Productivity or yield (kg/Bigha) ($X_{24}$) | 0.719** | 0.000 |
| 25. Gross return (Rs/ Bigha) ($X_{25}$) | -0.076 | 0.357 |
| 26. Utilization of cosmopolite sources of information($X_{26}$) | 0.054 | 0.508 |
| 27. Training received ($X_{27}$) | 0.025 | 0.758 |

*Correlation is significant at 5 % level, **Correlation is significant at 1 % level.

## Results

Table No 6.20 presents the correlation coefficient of consequent variable, marketed surplus ($Y_4$) with 27 independent variables.

It has been found that variables viz income (Rs /per capita/annum) ($X_8$), market Orientation ($X_{13}$), productivity or yield (kg/Bigha) ($X_{24}$) have recorded the positive correlation and attitude towards adoption ($X_{21}$). has recorded significant but negative correlation with marketed surplus ($Y_4$).

## Revelation

The result shows that income (Rs /per capita/annum) regress when farmers are producing ample of marketed surplus. This would suggest that, farmers need to build up better empirical skill for generating both income and marketed surplus together.

It also inventing that better market orientation, has contributed to generation of higher marketed surplus but attitude towards adoption has recorded but significant correlation. This would suggest that attitude toward adoption is not that relevant in generating marketed surplus.

Quite logical, productivity or yield has recorded a positive impact on marketed surplus.

**Model No 6.7. Model of Correlation coefficient of Marketed surplus ($Y_4$) with 27 independent variables**

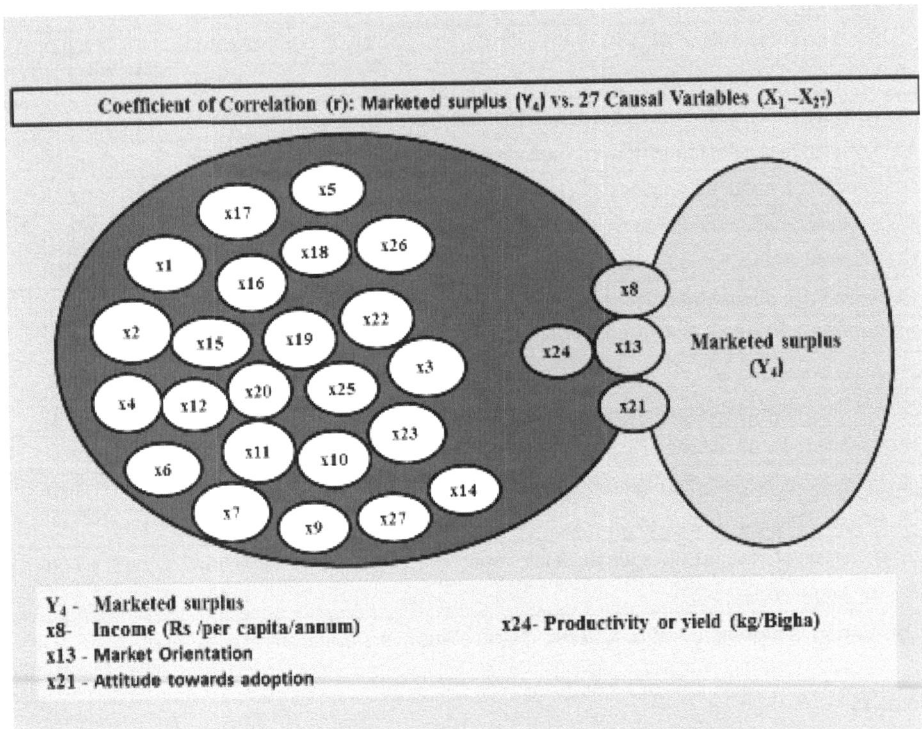


**Table No 6.21 Coefficients of Multiple regression analysis of Marketed surplus ($Y_4$) with 27 causal variables**

| Variables | Unstandardized Coefficients | | Standardized Coefficients | T | Sig. |
|---|---|---|---|---|---|
| | B | Std. Error | Beta | | |
| 1. Age ($X_1$) | -0.306 | 0.285 | -0.066 | -1.071 | 0.286 |
| 2. Education ($X_2$) | 0.363 | 1.208 | 0.019 | 0.301 | 0.764 |
| 3. Family size ($X_3$) | -0.051 | 0.714 | -0.005 | -0.071 | 0.943 |
| 4. Family education ($X_4$) | -0.379 | 1.426 | -0.017 | -0.266 | 0.791 |
| 5. Area under pulse cultivation (Bigha) ($X_5$) | -6.433 | 9.051 | -0.108 | -0.711 | 0.479 |
| 6. Farm size and technology adoption (Bigha) ($X_6$) | -1.140 | 0.766 | -0.129 | -1.489 | 0.139 |
| 7. No of crop diversity ($X_7$) | -3.780 | 7.804 | -0.041 | -0.484 | 0.629 |
| 8. Income (Rs /per capita/ annum) ($X_8$) | 6.910-005 | 0.000 | 0.032 | 0.510 | 0.611 |
| 9. Risk Orientation ($X_9$) | -1.930 | 2.352 | -0.051 | -0.821 | 0.413 |

| Variables | Unstandardized Coefficients | | Standardized Coefficients | T | Sig. |
|---|---|---|---|---|---|
| | B | Std. Error | Beta | | |
| 10. Scientific Orientation ($X_{10}$) | -3.665 | 2.406 | -0.099 | -1.523 | 0.130 |
| 11. Planning orientation ($X_{11}$) | -0.272 | 2.389 | -0.007 | -0.114 | 0.910 |
| 12. Production orientation ($X_{12}$) | 0.473 | 2.261 | 0.013 | 0.209 | 0.835 |
| 13. Market Orientation ($X_{13}$) | 5.513 | 2.968 | 0.118 | 1.858 | 0.066 |
| 14. Farmers attitude towards Pulse crop cultivation ($X_{14}$) | -7.420 | 3.270 | -0.160 | -2.269 | 0.025 |
| 15. Knowledge level of farmer towards cultivation of pulse crop ($X_{15}$) | 2.303 | 1.438 | 0.114 | 1.602 | 0.112 |
| 16. Knowledge about insecticides ($X_{16}$) | 7.857 | 13.028 | 0.037 | 0.603 | 0.548 |
| 17. Knowledge about fungicide ($X_{17}$) | 5.068 | 13.401 | 0.024 | 0.378 | 0.706 |
| 18. Knowledge about weed control ($X_{18}$) | 16.184 | 8.349 | 0.127 | 1.938 | 0.055 |
| 19. Knowledge about IPM practice ($X_{19}$) | -21.575 | 19.687 | -0.066 | -1.096 | 0.275 |
| 20. Farmers attitude towards IPM programme ($X_{20}$) | -5.158 | 3.672 | -0.087 | -1.405 | 0.163 |
| 21. Attitude towards adoption ($X_{21}$) | -5.760 | 3.267 | -0.111 | -1.763 | 0.080 |
| 22. Attitude towards discontinuous ($X_{22}$) | -1.909 | 2.811 | -0.043 | -0.679 | 0.498 |
| 23. Attitude towards rejection ($X_{23}$) | -9.273 | 5.264 | -0.109 | -1.762 | 0.081 |
| 24. Productivity or yield (kg/ Bigha) ($X_{24}$) | 0.676 | 0.061 | 0.730 | 11.168 | 0.000 |
| 25. Gross return (Rs/ Bigha) ($X_{25}$) | 0.001 | 0.001 | 0.146 | 0.993 | 0.323 |
| 26. Utilization of cosmopolite sources of information($X_{26}$) | 6.309 | 12.568 | 0.031 | 0.502 | 0.617 |
| 27. Training received ($X_{27}$) | 0.138 | 0.985 | 0.009 | 0.140 | 0.889 |

**Table No 6.22. Multiple regression analysis of Marketed surplus ($Y_4$) with 27 causal variables (Model Summary)**

| Model | R | R Square | Adjusted R Square | Std. Error of the Estimate | Change Statistics | | | | |
|---|---|---|---|---|---|---|---|---|---|
| | | | | | R Square Change | F Change | df1 | df2 | Sig. F Change |
| 1 | 0.778[a] | 0.605 | 0.517 | 30.835196 | 0.605 | 6.916 | 27 | 122 | 0.000 |

**Results**

Table No 6.21 Presents the multiple regression analysis which reflects the functional efficacy of the correlation through 'beta' value and respective 't' values of the casual variables on the consequent variables i.e marketed surplus ($Y_4$)

Table No 6.22 Presents The $R^2$ value 0.605 being, it is to conclude that even combination of all 27 variables so far, 60.5 % of variability embedded with the conseuent variable has been explained.

**Table No 6.23. Stepwise regression analysis of Marketed surplus ($Y_4$) with 27 causal variables**

| Model | | Unstandardized Coefficients | | Standardized Coefficients | t | Sig. |
|---|---|---|---|---|---|---|
| | | B | Std. Error | Beta | | |
| 1 | (Constant) | -10.643 | 18.009 | | -0.591 | 0.555 |
| | $X_{24}$ | 0.665 | 0.053 | 0.719 | 12.578 | 0.000 |

a. Dependent Variable: $Y_4$

**Table No 6.24. Model Summary of stepwise regression of Marketed surplus ($Y_4$) with 27 causal variables**

| Model | R | R Square | Adjusted R Square | Std. Error of the Estimate | Change Statistics | | | | |
|---|---|---|---|---|---|---|---|---|---|
| | | | | | R Square Change | F Change | df1 | df2 | Sig. F Change |
| 1 | 0.719[a] | 0.517 | 0.513 | 30.961528 | 0.517 | 158.215 | 1 | 148 | 0.000 |

**Results**

Table No 6.23 presents the step down multiple regression. It has been found that the variable productivity or yield (kg/Bigha) ($X_{24}$) have been retained after eliminating the trivial in the preceding step.

Table No 6.24 presents these variable explained 51.7 per cent of total '$r^2$'values of 60.5 per cent. So 1 variable of the total 27 variables merit highest importance in marketed surplus ($Y_4$).

**Revelation**

The variables retained in the last stage in stepwise regression analysis does present and operational constellation of 1 dominant variable working together and interacting ridiculously it can characterize both the level and direction of disillusionment.

So, these variable can go immensely important in making the farmers relinquished of disillusionment and thus have incubated an important strategic implementation for research locale and similes.

**Model No 6.8. Model of Stepwise regression analysis of Marketed surplus ($Y_4$) with 27 causal variables**

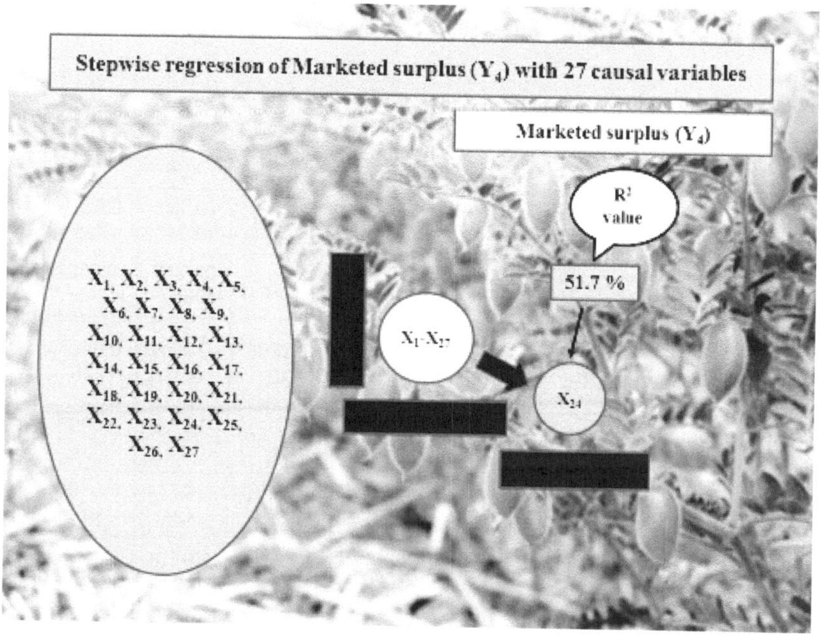


**Table No 6.25. Path analysis of comprehensive Marketed surplus ($Y_4$) with 27 causal variables**

| Variables | Total effect | Direct effect | Indirect effect | Substantial effect | | |
|---|---|---|---|---|---|---|
| | | | | I | II | III |
| 1. Age ($X_1$) | 0.0100 | -0.0657 | 0.0757 | 0.0562 X24 | 0.0182 X14 | 0.0103 X15 |
| 2. Education ($X_2$) | 0.0520 | 0.0201 | 0.0319 | 0.0358 X24 | 0.0186 X13 | -0.0132 X15 |
| 3. Family size ($X_3$) | -0.0540 | -0.0047 | -0.0493 | -0.0416 X24 | -0.0391 X5 | 0.0391 X25 |
| 4. Family education (X4) | 0.0240 | -0.0172 | 0.0412 | 0.0209 X13 | -0.0142 X15 | 0.0121 X21 |
| 5. Area under pulse cultivation (Bigha) ($X_5$) | -0.0920 | -0.1087 | 0.0167 | 0.1318 X25 | -0.0774 X14 | -0.0760 X24 |
| 6. Farm size and technology adoption (Bigha) ($X_6$) | -0.0520 | -0.1288 | 0.0768 | 0.0475 X24 | 0.0285 X7 | 0.0104 X13 |
| 7. Number of crop diversity ($X_7$) | 0.0260 | -0.0415 | 0.0675 | 0.0886 X6 | -0.0148 X15 | 0.0131 X24 |
| 8. Income (Rs /per capita/ annum) ($X_8$) | 0.2150 | 0.0326 | 0.1824 (III) | 0.1928 X24 | 0.0056 X18 | 0.0055 X1 |
| 9. Risk Orientation ($X_9$) | -0.0040 | -0.0521 | 0.0481 | 0.0308 X18 | 0.0307 X24 | -0.0170 X15 |

| Variables | Total effect | Direct effect | Indirect effect | Substantial effect | | |
|---|---|---|---|---|---|---|
| | | | | I | II | III |
| 10. Scientific Orientation ($X_{10}$) | -0.0090 | -0.0990 | 0.0900 | 0.0256 X24 | 0.0255 X15 | 0.0146 X23 |
| 11. Planning orientation ($X_{11}$) | 0.1320 | -0.0075 | 0.1395 | 0.1132 X24 | 0.0222 X18 | 0.0171 X6 |
| 12. Production orientation ($X_{12}$) | -0.0060 | 0.0127 | -0.0187 | -0.0402 X24 | 0.0120 X6 | -0.0111 X15 |
| 13. Market Orientation ($X_{13}$) | 0.1610 | 0.1179 | 0.0431 | 0.0745 X24 | 0.0218 X21 | -0.0113 X6 |
| 14. Farmers attitude towards Pulse crop cultivation ($X_{14}$) | -0.1200 | -0.1597 (II) | 0.0397 | 0.0740 X25 | -0.0527 X5 | 0.0287 X15 |
| 15. Knowledge level of farmer towards cultivation of pulse crop ($X_{15}$) | -0.1060 | 0.1140 | -0.2200 (II) | -0.1242 X24 | 0.0458 X25 | -0.0402 X14 |
| 16. Knowledge about insecticide ($X_{16}$) | 0.0010 | 0.0369 | -0.0359 | -0.0329 X24 | 0.0099 X6 | 0.0098 X23 |
| 17. Knowledge about fungicide ($X_{17}$) | -0.0170 | 0.0237 | -0.0407 | -0.0774 X24 | 0.0196 X18 | 0.0171 X15 |
| 18. Knowledge about weed control ($X_{18}$) | 0.1400 | 0.1274 | 0.0126 | 0.0716 X24 | -0.0184 X20 | -0.0148 X23 |
| 19. Knowledge about IPM practice ($X_{19}$) | 0.0190 | -0.0656 | 0.0846 | 0.0789 X24 | 0.0128 X15 | 0.0126 X25 |
| 20. Farmers attitude towards IPM programme ($X_{20}$) | -0.0340 | -0.0876 | 0.0536 | 0.0268 X18 | -0.0205 X24 | 0.0151 X21 |
| 21. Attitude towards adoption ($X_{21}$) | -0.1890 | -0.1110 | -0.0780 | -0.1037 X24 | -0.0231 X13 | 0.0180 X15 |
| 22. Attitude towards discontinuous ($X_{22}$) | 0.0490 | -0.0432 | 0.0922 | -0.0179 X22 | 0.0161 X14 | 0.0136 X10 |
| 23. Attitude towards rejection ($X_{23}$) | -0.0200 | -0.1087 | 0.0887 | 0.0621 X24 | 0.0173 X18 | 0.0133 X10 |
| 24. Productivity or yield (kg/ Bigha) ($X_{24}$) | 0.7190 | 0.7304 (I) | -0.0114 | -0.0194 X15 | 0.0158 X21 | -0.0143 X25 |
| 25. Gross return (Rs/ Bigha) ($X_{25}$) | -0.0760 | 0.1459 (III) | -0.2219 (I) | -0.0982 X5 | -0.0810 X14 | -0.0716 X24 |
| 26. Utilization/Information/ Cosmopolite ($X_{26}$) | 0.0540 | 0.0305 | 0.0235 | 0.0307 X24 | -0.0255 X15 | 0.0169 X6 |
| 27. Training received ($X_{27}$) | 0.0250 | 0.0250 | 0.0000 | 0.0433 X25 | -0.0327 X5 | -0.0286 X14 |

## Results

Table No 6.25 presents the path analysis, by decomposing the total effect (r) of antecedent variables into direct indirect effect and residual effect. Path analysis has been administered to get direction and network of influence of antecedent variables on consequent variable.

From the table, it's is clear that variable, productivity or yield (kg/Bigha) ($X_{24}$) has exerted highest direct effect on **marketed surplus ($Y_4$)** followed by farmer's attitude towards Pulse crop cultivation ($X_{14}$) and gross return (Rs/ Bigha) ($X_{25}$).

In case of indirect effect on **marketed surplus ($Y_4$)** gross return (Rs/ Bigha) ($X_{25}$) has exerted the highest indirect effect on **marketed surplus ($Y_4$)** followed by knowledge level of farmer towards cultivation of pulse crop ($X_{15}$) and income (Rs /per capita/annum) ($X_8$).

It is discernible from the table the highest number of variables (22) has routed their substantial indirect effect through the variable, productivity or yield (kg/ Bigha) ($X_{24}$). So, it could be inferred that these variables have got both substantive and associational properties to characterize the **marketed surplus ($Y_4$).**

The residual effect being 0.3948, it is to conclude that 39.48 per cent of variation in this relation could not be explained.

**Table No 6.26. Correlation coefficient of Marketable surplus ($Y_5$) with 27 independent variables**

| Variables | r- value | Sig. (2-tailed) |
|---|---|---|
| 1. Age ($X_1$) | 0.039 | 0.632 |
| 2. Education ($X_2$) | 0.170* | 0.038 |
| 3. Family size ($X_3$) | -0.185* | 0.023 |
| 4. Family education (X4) | 0.048 | 0.557 |
| 5. Area under pulse cultivation (Bigha) ($X_5$) | -0.114 | 0.165 |
| 6. Farm size and technology adoption (Bigha) ($X_6$) | 0.071 | 0.391 |
| 7. No of crop diversity ($X_7$) | 0.009 | 0.911 |
| 8. Income (Rs /per capita/annum) ($X_8$) | 0.203* | 0.013 |
| 9. Risk Orientation ($X_9$) | 0.037 | 0.656 |
| 10. Scientific Orientation ($X_{10}$) | 0.049 | 0.554 |
| 11. Planning orientation ($X_{11}$) | 0.144 | 0.079 |
| 12. Production orientation ($X_{12}$) | -0.030 | 0.719 |
| 13. Market Orientation ($X_{13}$) | 0.024 | 0.767 |
| 14. Farmers attitude towards Pulse crop cultivation ($X_{14}$) | -0.080 | 0.328 |
| 15. Knowledge level of farmer towards cultivation of pulse crop ($X_{15}$) | -0.121 | 0.140 |
| 16. Knowledge about insecticides ($X_{16}$) | -0.088 | 0.285 |
| 17. Knowledge about fungicide ($X_{17}$) | -0.056 | 0.498 |
| 18. Knowledge about weed control ($X_{18}$) | 0.084 | 0.306 |
| 19. Knowledge about IPM practice ($X_{19}$) | 0.097 | 0.239 |
| 20. Farmers attitude towards IPM programme ($X_{20}$) | -0.080 | 0.329 |
| 21. Attitude towards adoption ($X_{21}$) | -0.075 | 0.364 |
| 22. Attitude towards discontinuous ($X_{22}$) | 0.044 | 0.592 |
| 23. Attitude towards rejection ($X_{23}$) | 0.037 | 0.656 |

| Variables | r- value | Sig. (2-tailed) |
|---|---|---|
| 24. Productivity or yield (kg/Bigha) ($X_{24}$) | 0.843** | 0.000 |
| 25. Gross return (Rs/ Bigha) ($X_{25}$) | -0.103 | 0.208 |
| 26. Utilization of cosmopolite sources of information($X_{26}$) | 0.041 | 0.616 |
| 27. Training received ($X_{27}$) | -0.023 | 0.779 |

*Correlation is significant at 5 % level **Correlation is significant at 1 % level

## Results

Table No 6.26. presents the correlation coefficient of consequent variable, marketable surplus ($Y_5$) with 27 independent variables.

It has been found that variables viz education ($X_2$), income (rs /per capita/ annum) ($X_8$), productivity or yield (kg/bigha) ($X_{24}$) have recorded the positive and significant correlation and family size ($X_3$) has recorded significant but negative correlation with marketable surplus ($Y_5$).

## Revelation

It is interesting to note that for small size of family generation of marketable surplus is more and those, who have got higher level of education they are generating more marketable surplus. Productivity and income both have got positive impact on marketable surplus.

**Model No 6.9 Model of Correlation coefficient of Marketable surplus ($Y_5$) with 27 independent variables**

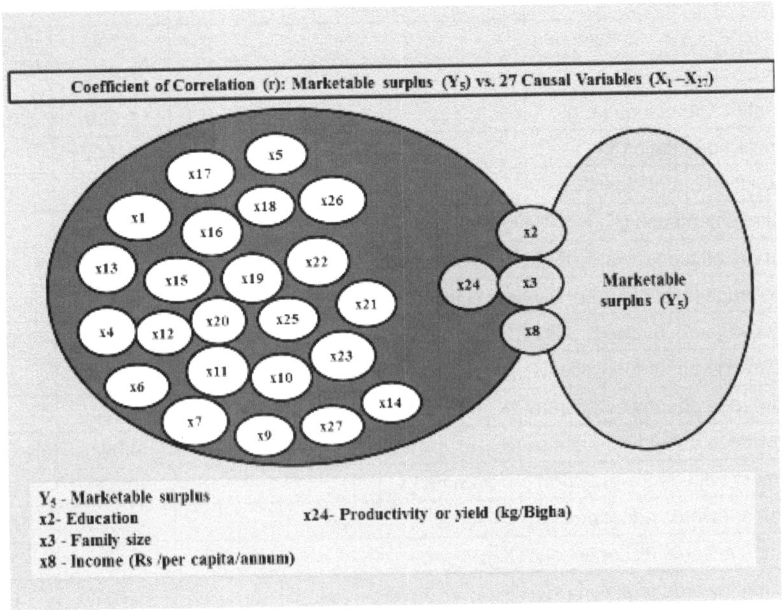

**Table No 6.27. Coefficients of Multiple regression analysis of Marketable surplus ($Y_5$) with 27 causal variables**

| Variables | Unstandardized Coefficients | | Standardized Coefficients | T | Sig. |
|---|---|---|---|---|---|
| | B | Std. Error | Beta | | |
| 1. Age ($X_1$) | -0.068 | 0.188 | -0.017 | -0.362 | 0.718 |
| 2. Education ($X_2$) | 2.243 | 0.796 | 0.138 | 2.818 | 0.006 |
| 3. Family size ($X_3$) | -1.400 | 0.471 | -0.152 | -2.976 | 0.004 |
| 4. Family education ($X_4$) | 0.459 | 0.939 | 0.024 | 0.489 | 0.626 |
| 5. Area under pulse cultivation (Bigha) ($X_5$) | 5.901 | 5.962 | 0.114 | 0.990 | 0.324 |
| 6. Farm size and technology adoption (Bigha) ($X_6$) | 0.491 | 0.504 | 0.064 | 0.974 | 0.332 |
| 7. No of crop diversity ($X_7$) | 5.582 | 5.140 | 0.070 | 1.086 | 0.280 |
| 8. Income (Rs /per capita/ annum) ($X_8$) | -1.241-005 | 0.000 | -0.007 | -0.139 | 0.890 |
| 9. Risk Orientation ($X_9$) | 0.691 | 1.549 | 0.021 | 0.446 | 0.656 |
| 10. Scientific Orientation ($X_{10}$) | -1.089 | 1.585 | -0.034 | -0.687 | 0.493 |
| 11. Planning orientation ($X_{11}$) | 0.293 | 1.574 | 0.009 | 0.186 | 0.853 |
| 12. Production orientation ($X_{12}$) | 0.919 | 1.489 | 0.028 | 0.617 | 0.538 |
| 13. Market Orientation ($X_{13}$) | -2.822 | 1.955 | -0.069 | -1.444 | 0.151 |
| 14. Farmers attitude towards Pulse crop cultivation ($X_{14}$) | -5.067 | 2.154 | -0.125 | -2.352 | 0.020 |
| 15. Knowledge level of farmer towards cultivation of pulse crop ($X_{15}$) | 1.122 | 0.947 | 0.064 | 1.185 | 0.238 |
| 16. Knowledge about insecticides ($X_{16}$) | -8.162 | 8.581 | -0.044 | -0.951 | 0.343 |
| 17. Knowledge about fungicide ($X_{17}$) | 7.562 | 8.827 | 0.040 | 0.857 | 0.393 |
| 18. Knowledge about weed control ($X_{18}$) | 0.555 | 5.499 | 0.005 | 0.101 | 0.920 |
| 19. Knowledge about IPM practice ($X_{19}$) | 8.921 | 12.967 | 0.031 | 0.688 | 0.493 |
| 20. Farmers attitude towards IPM programme ($X_{20}$) | -3.366 | 2.419 | -0.065 | -1.392 | 0.167 |
| 21. Attitude towards adoption ($X_{21}$) | 1.489 | 2.152 | 0.033 | 0.692 | 0.490 |
| 22. Attitude towards discontinuous ($X_{22}$) | -1.462 | 1.852 | -0.038 | -0.790 | 0.431 |
| 23. Attitude towards rejection ($X_{23}$) | -2.986 | 3.467 | -0.040 | -0.861 | 0.391 |
| 24. Productivity or yield (kg/ Bigha) ($X_{24}$) | 0.691 | 0.040 | 0.857 | 17.333 | 0.000 |

| Variables | Unstandardized Coefficients | | Standardized Coefficients | T | Sig. |
|---|---|---|---|---|---|
| | B | Std. Error | Beta | | |
| 25. Gross return (Rs/ Bigha) ($X_{25}$) | 0.000 | 0.001 | -0.023 | -0.205 | 0.838 |
| 26. Utilization of cosmopolite sources of information($X_{26}$) | 3.314 | 8.278 | 0.019 | 0.400 | 0.690 |
| 27. Training received ($X_{27}$) | -0.407 | 0.649 | -0.030 | -0.628 | 0.531 |

**Table No 6.28 Multiple regression analysis of Marketable surplus ($Y_5$) with 27 causal variables (Model Summary)**

| Model | R | R Square | Adjusted R Square | Std. Error of the Estimate | Change Statistics | | | | |
|---|---|---|---|---|---|---|---|---|---|
| | | | | | R Square Change | F Change | df1 | df2 | Sig. F Change |
| 1 | 0.880[a] | 0.774 | 0.724 | 20.310141 | 0.774 | 15.472 | 27 | 122 | 0.000 |

**Result**

Table No 6.27 presents the multiple regression analysis which reflects the functional efficacy of the correlation through 'beta' value and respective 't' values of the casual variables on the consequent variables i.e marketable surplus ($Y_5$)

Table No 6.28 the $R^2$ value 0.774 being, it is to conclude that even combination of all 27 variables so far, 77.4 % of variability embedded with the consequent variable has been explained.

**Table No 6.29 Stepwise regression analysis of Marketable surplus ($Y_5$) with 27 causal variables**

| Model | | Unstandardized Coefficients | | Standardized Coefficients | t | Sig. |
|---|---|---|---|---|---|---|
| | | B | Std. Error | Beta | | |
| 1 | (Constant) | 6.725 | 12.150 | | 0.553 | 0.581 |
| | $X_{24}$ | 0.679 | 0.036 | 0.843 | 19.034 | 0.000 |
| 2 | (Constant) | 17.134 | 12.227 | | 1.401 | 0.163 |
| | $X_{24}$ | 0.673 | 0.035 | 0.835 | 19.405 | 0.000 |
| | $X_3$ | -1.273 | 0.397 | -0.138 | -3.203 | 0.002 |
| 3 | (Constant) | 8.749 | 12.306 | | 0.711 | 0.478 |
| | $X_{24}$ | 0.669 | 0.034 | 0.829 | 19.716 | 0.000 |
| | $X_3$ | -1.190 | 0.389 | -0.129 | -3.057 | 0.003 |
| | $X_2$ | 1.938 | 0.684 | 0.119 | 2.832 | 0.005 |

a. Dependent Variable: $Y_5$

**Table No 6.30 Model Summary of stepwise regression of Marketable surplus ($Y_5$) with 27 causal variables**

| Model | R | R Square | Adjusted R Square | Std. Error of the Estimate | Change Statistics | | | | |
|---|---|---|---|---|---|---|---|---|---|
| | | | | | R Square Change | F Change | df1 | df2 | Sig. F Change |
| 1 | 0.843[a] | 0.710 | 0.708 | 20.888406 | 0.710 | 362.274 | 1 | 148 | 0.000 |
| 2 | 0.854[b] | 0.729 | 0.725 | 20.264152 | 0.019 | 10.259 | 1 | 147 | 0.002 |
| 3 | 0.862[c] | 0.743 | 0.738 | 19.796990 | 0.014 | 8.020 | 1 | 146 | 0.005 |

## Results

Table no 6.29 presents the step down multiple regression. It has been found that the variables productivity or yield (kg/Bigha) ($X_{24}$), family size ($X_3$), education ($X_2$) have been retained after eliminating the trivial in the preceding step.

Table no 6.30 presents these variables together have explained 74.3 per cent of total '$r^2$'values of 77.4 per cent. So 3 variables of the total 27 variables merit highest importance in marketable surplus ($Y_5$).

## Revelation

The variables retained in the last stage in stepwise regression analysis does present and operational constellation of 3 dominant variables working together and interacting ridiculously it can characterize both the level and direction of disillusionment.

So, those few variables can go immensely important in making the farmers relinquished of disillusionment and thus have incubated an important strategic implementation for research locale and similes.

**Model No 6.10 Model of Stepwise regression analysis of Marketable surplus ($Y_5$) with 27 causal variables**

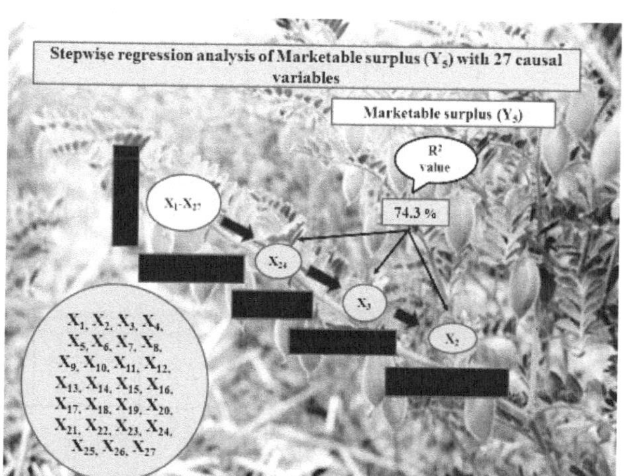

**Table No 6.31 Path analysis of comprehensive Marketable surplus ($Y_5$) with 27 causal variables**

| Variables | Total effect | Direct effect | Indirect effect | Substantial effect | | |
|---|---|---|---|---|---|---|
| | | | | I | II | III |
| 1. Age ($X_1$) | 0.0390 | -0.0181 | 0.0571 | 0.0660 X24 | -0.0182 X2 | 0.0143 X14 |
| 2. Education ($X_2$) | 0.1700 | 0.1379 (III) | 0.0321 | 0.0420 X24 | 0.0117 X3 | -0.0111 X13 |
| 3. Family size ($X_3$) | -0.1850 | -0.1505 (II) | -0.0345 | -0.0489 X24 | 0.0392 X5 | -0.0108 X2 |
| 4. Family education (X4) | 0.0480 | 0.0231 | 0.0249 | 0.0476 X2 | -0.0124 X13 | 0.0111 X24 |
| 5. Area under pulse cultivation (Bigha) ($X_5$) | -0.1140 | 0.1089 | -0.2229 (I) | -0.0892 X24 | -0.0609 X14 | -0.0542 X3 |
| 6. Farm size and technology adoption (Bigha) ($X_6$) | 0.0710 | 0.0634 | 0.0076 | 0.0557 X24 | -0.0472 X7 | 0.0117 X2 |
| 7. Number of crop diversity ($X_7$) | 0.0090 | 0.0686 | -0.0596 | -0.0436 X6 | 0.0154 X24 | -0.0096 X5 |
| 8. Income (Rs /per capita/ annum) ($X_8$) | 0.2030 | -0.0063 | 0.2093 (II) | 0.2264 X24 | -0.0266 X3 | 0.0056 X16 |
| 9. Risk Orientation ($X_9$) | 0.0370 | 0.0209 | 0.0161 | 0.0360 X24 | -0.0132 X5 | -0.0095 X15 |
| 10. Scientific Orientation ($X_{10}$) | 0.0490 | -0.0331 | 0.0821 | 0.0283 X3 | 0.0143 X15 | 0.0121 X2 |
| 11. Planning orientation ($X_{11}$) | 0.1440 | 0.0086 | 0.1354 | 0.1329 X24 | -0.0084 X6 | 0.0080 X7 |
| 12. Production orientation ($X_{12}$) | -0.0300 | 0.0278 | -0.0578 | -0.0472 X24 | -0.0062 X15 | -0.0059 X6 |
| 13. Market Orientation ($X_{13}$) | 0.0240 | -0.0702 | 0.0942 | 0.0875 X24 | 0.0218 X2 | -0.0064 x21 |
| 14. Farmers attitude towards Pulse crop cultivation ($X_{14}$) | -0.0800 | -0.1256 | 0.0456 | 0.0528 X5 | 0.0161 X15 | -0.0120 X24 |
| 15. Knowledge level of farmer towards cultivation of pulse crop ($X_{15}$) | -0.1210 | 0.0640 | -0.1850 (III) | -0.1458 X24 | 0.0640 X15 | 0.0366 X5 |
| 16. Knowledge about insecticide ($X_{16}$) | -0.0880 | -0.0435 | -0.0445 | -0.0386 X24 | -0.0060 X3 | 0.0056 X7 |
| 17. Knowledge about fungicide ($X_{17}$) | -0.0560 | 0.0391 | -0.0951 | -0.0909 X24 | -0.0098 X2 | 0.0096 X15 |

| Variables | Total effect | Direct effect | Indirect effect | Substantial effect | | |
|---|---|---|---|---|---|---|
| | | | | I | II | III |
| 18. Knowledge about weed control ($X_{18}$) | 0.0840 | 0.0057 | 0.0783 | 0.0840 X24 | -0.0137 X20 | -0.0062 X15 |
| 19. Knowledge about IPM practice ($X_{19}$) | 0.0970 | 0.0321 | 0.0649 | 0.0926 X24 | 0.0103 X5 | -0.0088 X14 |
| 20. Farmers attitude towards IPM programme ($X_{20}$) | -0.0800 | -0.0654 | -0.0146 | -0.0240 X24 | -0.0127 X2 | 0.0113 X14 |
| 21. Attitude towards adoption ($X_{21}$) | -0.0750 | 0.0325 | -0.1075 | -0.1218 X21 | 0.0138 X13 | 0.0101 X15 |
| 22. Attitude towards discontinuous ($X_{22}$) | 0.0440 | -0.0375 | 0.0815 | 0.0557 X24 | -0.0141 X14 | 0.0134 X2 |
| 23. Attitude towards rejection ($X_{23}$) | 0.0370 | -0.0405 | 0.0775 | 0.0729 X24 | -0.0132 X3 | -0.0089 X5 |
| 24. Productivity or yield (kg/ Bigha) ($X_{24}$) | 0.8430 | 0.8575 (I) | -0.0145 | -0.0113 X5 | -0.0109 X15 | 0.0086 X3 |
| 25. Gross return (Rs/ Bigha) ($X_{25}$) | -0.1030 | -0.0183 | -0.0847 | 0.0983 X5 | -0.0840 X24 | -0.0637 X14 |
| 26. Utilization/Information/ Cosmopolite ($X_{26}$) | 0.0410 | 0.0178 | 0.0232 | 0.0360 X26 | -0.0143 X15 | 0.0090 X13 |
| 27. Training received ($X_{27}$) | -0.0230 | -0.0287 | 0.0057 | 0.0328 X5 | -0.0225 X14 | -0.0185 X3 |

## Results

Table No 6.31 presents the path analysis, by decomposing the total effect (r) of antecedent variables into direct indirect effect and residual effect. Path analysis has been administered to get direction and network of influence of antecedent variables on consequent variable.

From the table, it's is clear that variable, productivity or yield (kg/Bigha) ($X_{24}$) has exerted highest direct effect on **marketable surplus ($Y_5$)** followed by Family size ($X_3$) and education ($X_2$).

In case of indirect effect on **marketable surplus ($Y_5$)** area under pulse cultivation (Bigha) ($X_5$) has exerted the highest indirect effect on **marketable surplus ($Y_5$)** followed by Income (Rs /per capita/annum) ($X_8$) and knowledge level of farmer towards cultivation of pulse crop ($X_{15}$).

It is discernible from the table the highest number of variables (22) has routed their substantial indirect effect through the variable, productivity or yield (kg/ Bigha) ($X_{24}$). So, it could be inferred that these variables have got both substantive and associational properties to characterize the **marketable surplus ($Y_5$).**

The residual effect being 0.2263, it is to conclude that 22.63 per cent of variation in this relation could not be explained.

**Table No 6.32. Factor analysis: Conglomeration of 27 explanatory variables into 7 factors**

| Sl. No. | Variables accommodated with respective factor loading | Eigen values | % Variation | Cumulative | Factor Naming |
|---|---|---|---|---|---|
| Factor-1 | Area under pulse cultivation (Bigha) ($X_5$) (0.899)<br>Gross return (Rs/ Bigha) ($X_{25}$) (0.880)<br>Farmers attitude towards Pulse crop cultivation ($X_{14}$) (0.631)<br>Knowledge level of farmer towards cultivation of pulse crop ($X_{15}$) (0.574)<br>Training received ($X_{27}$) (0.439)<br>Family size ($X_3$) (0.329) | 2.879 | 10.662 | 10.662 | **Cognitive motivational** |
| Factor-2 | Farm size and technology adoption (Bigha) ($X_6$) (0.811)<br>No of crop diversity ($X_7$) (-0.700)<br>Family education ($X_4$) (0.383)<br>Market Orientation ($X_{13}$) (0.335)<br>Education ($X_2$) (0.333)<br>Planning orientation ($X_{11}$) (-0.292)<br>Attitude towards discontinuous ($X_{22}$) (0.288)<br>Utilization of cosmopolite sources of information($X_{26}$) (-0.212) | 1.918 | 7.102 | 17.764 | **Resource proficiency** |
| Factor-3 | Productivity or yield (kg/Bigha) ($X_{24}$) (0.306)<br>Age ($X_1$) (-0.254)<br>Farmers attitude towards IPM programme ($X_{20}$) (-0.245)<br>Knowledge about fungicide ($X_{17}$) (-0.221)<br>Income (Rs /per capita/annum) ($X_8$) (0.207) | 1.779 | 6.587 | 24.352 | **Economic performance** |
| Factor-4 | Knowledge about weed control ($X_{18}$) (0.585)<br>Attitude towards adoption ($X_{21}$) (-0.346)<br>Attitude towards rejection ($X_{23}$) (0.343) | 1.660 | 6.148 | 30.500 | **Socialization level** |
| Factor-5 | Scientific Orientation ($X_{10}$) (0.722)<br>Risk Orientation ($X_9$) (-0.252) | 1.538 | 5.695 | 36.195 | **Risk management** |
| Factor-6 | Knowledge about IPM practice ($X_{19}$) (0.380) | 1.377 | 5.098 | 41.294 | |
| Factor-7 | Knowledge about insecticides ($X_{16}$) (0.643)<br>Production orientation ($X_{12}$) (-0.251) | 1.370 | 5.075 | 46.369 | **Capacity level** |

## Results and Revelation

Table 6.32 presents factor analysis for conglomeration of apparently different variables into homogenous cluster called Factor and this conglomeration has been done based on a Factor Loading Eigen value So, the higher the value of factor loading the more is the "Imbibing Propensity" to bring homogenous variables to the closest binding.

Here the 27 variables have been accommodated in 7 Factors.

### FACTOR 1

FACTOR 1 has accommodated area under pulse cultivation (Bigha) ($X_5$), gross return (Rs/ Bigha) ($X_{25}$), farmer's attitude towards pulse crop cultivation ($X_{14}$), knowledge level of farmer towards cultivation of pulse crop ($X_{15}$), training received ($X_{27}$), family size ($X_3$) which contributed to 10.662 % and has been renamed as **Cognitive motivational.**

### FACTOR-2

FACTOR 2 has accommodated farm size and technology adoption (Bigha) ($X_6$), number of crop diversity ($X_7$), family education ($X_4$), market Orientation ($X_{13}$), education ($X_2$), planning orientation ($X_{11}$) , attitude towards discontinuous ($X_{22}$) ,utilization of cosmopolite sources of information($X_{26}$) which contributed to 7.102 % and has been renamed as **Resource proficiency.**

### FACTOR-3

FACTOR 3 has accommodated productivity or yield (kg/Bigha) ($X_{24}$), age ($X_1$), farmer's attitude towards IPM programme ($X_{20}$), knowledge about fungicide ($X_{17}$), income (Rs /per capita/annum) ($X_8$) which contributed to 6.587 % and has been renamed as **Economic performance.**

### FACTOR-4

FACTOR 4 has accommodated knowledge about weed control ($X_{18}$) (0.585), attitude towards adoption ($X_{21}$), attitude towards rejection ($X_{23}$) which contributed to 6.148 % and has been renamed as **Socialization level.**

### FACTOR-5

FACTOR 5 has accommodated scientific orientation ($X_{10}$), risk Orientation ($X_9$) which contributed to 5.695 % and has been renamed as **Risk management.**

### FACTOR-6

FACTOR 6 has accommodated knowledge about IPM practice ($X_{19}$) which contributed to 5.098 %.

### FACTOR-7

FACTOR 7 has accommodated knowledge about insecticides ($X_{16}$), production orientation ($X_{12}$) which contributed to 5.075 % and has been renamed as **Capacity level.**

**Table No 6.33 Canonical correlation Analysis to derive the interaction pattern of dependent variables**

| Dependent variable | Coefficients |
|---|---|
| 1. Socialization of pulse enterprise ($Y_1$) | -0.496 |
| 2. Socialization through replacement level ($Y_2$) | -0.349 |
| 3. Socialization through reinvention ($Y_3$) | -0.051 |
| 4. Marketed surplus ($Y_4$) | 0.173 |
| 5. Marketable surplus ($Y_5$) | 0.460 |

**Table No 6.34 Canonical correlation Analysis to derive the interaction pattern of independent variables**

| | Independent variable | Coefficients |
|---|---|---|
| 1. | Age ($X_1$) | -0.015 |
| 2. | Education ($X_2$) | 0.078 |
| 3. | Family size ($X_3$) | -0.029 |
| 4. | Family education ($X_4$) | -0.030 |
| 5. | Area under pulse cultivation (Bigha) ($X_5$) | -0.293 |
| 6. | Farm size and technology adoption (Bigha) ($X_6$) | 0.200 |
| 7. | No of crop diversity ($X_7$) | -0.005 |
| 8. | Income (Rs /per capita/annum) ($X_8$) | -0.044 |
| 9. | Risk Orientation ($X_9$) | 0.023 |
| 10. | Scientific Orientation ($X_{10}$) | 0.002 |
| 11. | Planning orientation ($X_{11}$) | -0.016 |
| 12. | Production orientation ($X_{12}$) | -0.000 |
| 13. | Market Orientation ($X_{13}$) | -0.068 |
| 14. | Farmers attitude towards Pulse crop cultivation ($X_{14}$) | -0.170 |
| 15. | Knowledge level of farmer towards cultivation of pulse crop ($X_{15}$) | 0.016 |
| 16. | Knowledge about insecticides ($X_{16}$) | -0.024 |
| 17. | Knowledge about fungicide ($X_{17}$) | 0.012 |
| 18. | Knowledge about weed control ($X_{18}$) | 0.001 |
| 19. | Knowledge about IPM practice ($X_{19}$) | -0.036 |
| 20. | Farmers attitude towards IPM programme ($X_{20}$) | -0.022 |
| 21. | Attitude towards adoption ($X_{21}$) | -0.029 |
| 22. | Attitude towards discontinuous ($X_{22}$) | -0.049 |
| 23. | Attitude towards rejection ($X_{23}$) | -0.060 |
| 24. | Productivity or yield (kg/Bigha) ($X_{24}$) | 0.661 |

| | | |
|---|---|---|
| 25. | Gross return (Rs/ Bigha) ($X_{25}$) | -0.249 |
| 26. | Utilization/Information/Cosmopolite ($X_{26}$) | -0.017 |
| 27. | Training received ($X_{27}$) | -0.076 |

**Model No 6.11 Model of Canonical covariate analysis: Interaction between left side (dependent) and right side (independent) variables**

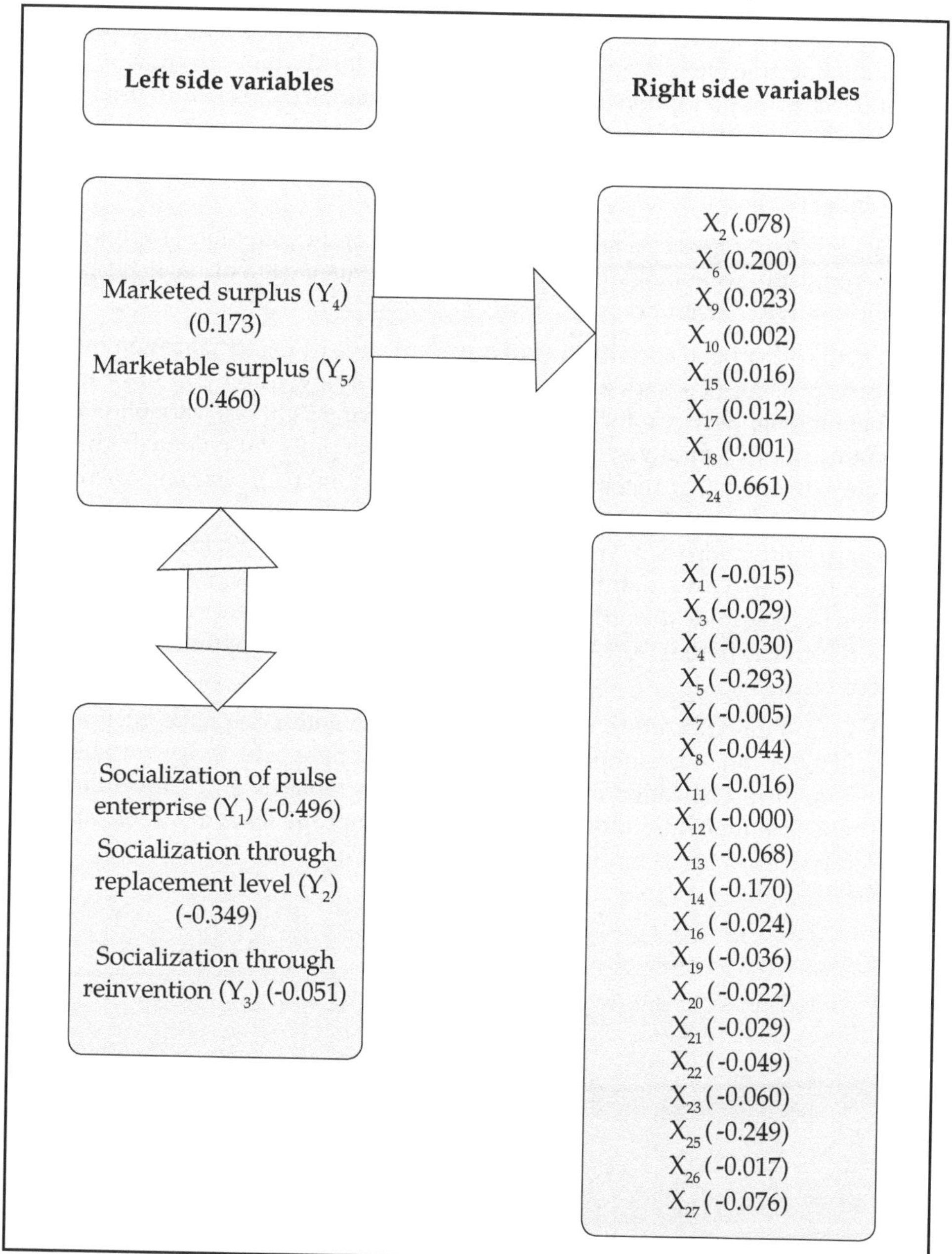

## Results

Table No 6.33 and 6.34 presents, canonical covariates analysis for this study has been carried out to extract the canonical covariates between two sets of variables, the left side variables and right side variables. it has offered us a dyadic tic interaction between left side variables variable's and right side variables.

It has further gone to form a congenital in group between marketed surplus ($Y_4$) and marketable surplus ($Y_5$) to imply that these two dependent variables have been in clandestine interaction with right side variables viz. the education ($X_2$), farm size and technology adoption (bigha) ($X_6$), risk orientation ($X_9$), scientific orientation ($X_{10}$), knowledge level of farmer towards cultivation of pulse crop ($X_{15}$), knowledge about fungicide ($X_{17}$), knowledge about weed control ($X_{18}$), productivity or yield (kg/ bigha) ($X_{24}$).

So, in order to generate marketed surplus ($Y_4$) and marketable surplus ($Y_5$) in pulse crop, land, technology, knowledge and risk management have got precise and substantive contribution.

On the other hand socialization of pulse enterprise ($Y_1$), socialization through replacement level ($Y_2$) and socialization through reinvention ($Y_3$) have moved together for a clandestine interaction with the following right side variables viz. age ($X_1$), family size ($X_3$), family education ($X_4$), area under pulse cultivation (bigha) ($X_5$), no of crop diversity ($X_7$), income (rs/per capita/annum) ($X_8$), planning orientation ($X_{11}$), production orientation ($X_{12}$), market orientation ($X_{13}$) ,farmers attitude towards pulse crop cultivation ($X_{14}$), knowledge about insecticides ($X_{16}$), knowledge about IPM practice ($X_{19}$), farmers attitude towards IPM programme ($X_{20}$), attitude towards adoption ($X_{21}$), attitude towards discontinuous ($X_{22}$),attitude towards rejection ($X_{23}$) , gross return (rs/bigha) ($X_{25}$), utilization/information/cosmopolite ($X_{26}$), training received ($X_{27}$).

So, all the socialization components in pulse enterprise have been able to form a "Holon" and have impacted on the selected right side causal variable's. In conclusion, the socialization in pulse crop imbibes replacement, reinvention and entrepreneurial innovation together to form a composite level of socialization in pulse crop.

# 6
# CHAPTER

# SUMMARY AND CONCLUSION

The present study has been designed to estimate the technology socialization process and market behaviour of pulse enterprise status between two villages, Maheswarpur and Panchkahoniya from West Bengal. It has been done so with a presumption that the villages Maheswarpur and Panchkahoniya has already gone a profile of modernization process increasingly exposed to stimulus of modernization in the form of ratification method, rural communication, and a good history of clustered irrigation system, wider exposure to supply and agricultural marketing system and a blessed proximity to agricultural university.

With this background the general objective has been carved as follows:

And the following specific objectives are drawn up to organize the entire space and pace of study under the entitlement of

"Technology Socialization Process and Market Behaviour of Pulse Enterprise: The Structural and Functional Analysis"

1. To generate basic information's on socialization of pulse crop in the study area.
2. To identify and standardize the variables, dependent and independent, impacting on both socialization and market behaviour of pulse crop in the study area.
3. To elucidate the inter and intra level interaction between dependent variables.i.e. socialization and market behaviour. with those of selected socio economic and ecological variables.
4. To organize participatory analysis on prospects and constraints of pulse enterprises in the study area in term of its socialization and market behaviour.
5. To delineate in micro level policy Based on the empirical result on effective socialization process and market behaviour of pulse crop.

## Methodologies

The present study was conducted in West Bengal. village Maheswarpur of Chakdah block and Panchkahoniya of Haringhata block in Nadia district of West Bengal were selected purposively for the study. Face to Face interview was conducted with the help of structured interview schedule. Total number of respondents was 150. Out 150, 75 respondents were selected with the help of simple random sample method from the village, Maheswarpur and another 75 respondents were selected with the help of simple random sample method from the village Panchkahoniya.

## The independent variables

Age ($X_1$), Education ($X_2$), Family size ($X_3$), Family education ($X_4$), Area under pulse cultivation (Bigha) ($X_5$), Farm size and technology adoption (Bigha) ($X_6$), No of crop diversity ($X_7$), Income (Rs /per capita/annum) ($X_8$), Risk Orientation ($X_9$), Scientific Orientation ($X_{10}$), Planning orientation ($X_{11}$), Production orientation ($X_{12}$), Market Orientation ($X_{13}$), Farmers attitude towards Pulse crop cultivation ($X_{14}$), Knowledge level of farmer towards cultivation of pulse crop ($X_{15}$), Knowledge about insecticides ($X_{16}$), Knowledge about fungicide ($X_{17}$), Knowledge about weed control ($X_{18}$), Knowledge about IPM practice ($X_{19}$), Farmers attitude towards IPM programme ($X_{20}$), Attitude towards adoption ($X_{21}$), Attitude towards discontinuous ($X_{22}$), Attitude towards rejection ($X_{23}$), Productivity or yield (kg/Bigha) ($X_{24}$), Gross return (Rs/ Bigha) ($X_{25}$), Utilization of cosmopolite sources of information ($X_{26}$), Training received ($X_{27}$).

## The Dependent variables

Socialization of pulse enterprise ($Y_1$), Socialization through replacement level ($Y_2$), Socialization through reinvention ($Y_3$), Marketed surplus ($Y_4$), Marketable surplus ($Y_5$).

## Statistical tools

Following statistical tools were used for the analysis

1. Mean
2. Standard deviation
3. Coefficient of Variance
4. Correlation of coefficient
5. Multiple regression analysis
6. Factor analysis
7. Path analysis
8. Canonical Correlation Analysis

## FINDINGS

- **Socialization of pulse enterprise ($Y_1$)**

### Coefficient of Correlation

The following independent variables *viz* family size ($X_3$), area under pulse cultivation (bigha) ($X_5$), farmer's attitude towards pulse crop cultivation ($X_{14}$), knowledge level of farmer towards cultivation of pulse crop ($X_{15}$), gross return (rs/ bigha) ($X_{25}$), training received ($X_{27}$) have recorded the positive and productivity or yield (kg/ bigha) ($X_{24}$) has recorded significant but negative correlation with socialization of pulse enterprise.

### Multiple regression analysis

The $R^2$ value 0.787 being, it's is to conclude that even a combination of all 27 variables so far, 78.7 % of variability embedded with the consequent variable has been explained.

### Step down multiple regression analysis

The following causal variables *viz.* gross return (Rs/ Bigha) ($X_{25}$), area under pulse cultivation (Bigha) ($X_5$), training Received ($X_{27}$), productivity or yield (Kg/ Bigha) ($X_{24}$), farmer's attitude towards pulse crop cultivation ($X_{14}$) have been retained after eliminating the trivial in the preceding step.

So, these variables together have explained 76.8 per cent of total 'R' values of 78.7 per cent. So 5 variables of the total 27 variables merit highest importance in socialization of pulse enterprise ($Y_1$).

### Path Analysis

The exogenous variables, gross return (Rs/ Bigha) ($X_{25}$) has exerted highest direct effect on **socialization of pulse enterprise ($Y_1$)** followed by area under pulse cultivation (Bigha) ($X_5$) and productivity or yield (kg/Bigha) ($X_{24}$).

In case of indirect effect on **socialization of pulse enterprise ($Y_1$)** followed by gross return (Rs/ Bigha) ($X_{25}$) and farmer's attitude towards pulse crop cultivation ($X_{14}$).

It is discernible from the table the highest number of variables (21) has routed their substantial indirect effect through the variable, area under pulse cultivation (Bigha) ($X_5$). So, it could be inferred that these variables have got both substantive and associational properties to characterize the **socialization of pulse enterprise ($Y_1$).**

The residual effect being 0.2124, it is to conclude that 21.24 per cent of variation in this relation could not be explained.

- **Socialization through replacement level ($Y_2$)**

### Coefficient of Correlation

The following independent variables *viz.* area under pulse cultivation (bigha) ($X_5$), no of crop diversity ($X_7$), farmer's attitude towards pulse crop cultivation

($X_{14}$), gross return (rs/ bigha) ($X_{25}$), have recorded the positive and farm size and technology adoption (bigha) ($X_6$), risk orientation ($X_9$) has recorded significant but negative correlation with socialization of replacement level ($Y_2$).

### Multiple regression analysis

The $R^2$ value 0.689 being, it's is to conclude that even combination of all 27 variables so far, 68.9 % of variability embedded with the consequent variable has been explained.

### Step down multiple regression analysis

The following causal variables *viz* farm size and technology adoption (bigha) ($X_6$), area under pulse cultivation (bigha) ($X_5$), attitude towards discontinuous ($X_{22}$), knowledge about IPM practice ($X_{19}$) have been retained after eliminating the trivial in the preceding step.

So these variables together have explained 62.1 per cent of total '$r^2$' values of 68.9 per cent. So 4 variables of the total 27 variables merit highest importance in socialization through replacement level ($Y_2$).

### Path Analysis

The exogenous variables, farm size and technology adoption (Bigha) ($X_6$) has exerted highest direct effect on **socialization through replacement level ($Y_2$)** followed by area under pulse cultivation (Bigha) ($X_5$) and knowledge about IPM practice ($X_{19}$).

In case of indirect effect on **socialization through replacement level ($Y_2$)** area under pulse cultivation (Bigha) ($X_5$) has exerted the highest indirect effect on **socialization through replacement level ($Y_2$)** followed by number of crop diversity ($X_7$) and farmer's attitude towards pulse crop cultivation ($X_{14}$).

It is discernible from the table the highest number of variables (20) has routed their substantial indirect effect through the variable, Farm size and technology adoption (Bigha) ($X_6$). So, it could be inferred that these variables have got both substantive and associational properties to characterize the **socialization through replacement level ($Y_2$).**

The residual effect being 0.3102, it is to conclude that 31.02 per cent of variation in this relation could not be explained.

## • Socialization through reinvention ($Y_3$)

### Coefficient of Correlation

The following independent variables *viz.* attitude towards adoption ($X_{21}$), training received ($X_{27}$) have recorded the positive and significant correlation with socialization through reinvention ($Y_3$).

### Multiple regression analysis

The $R^2$ value 0.381 being, it's is to conclude that even combination of all 27 variables so far, 38.1 % of variability embedded with the consequent variable has been explained.

**Step down multiple regression analysis**

The following causal variables *viz* training received ($X_{27}$), attitude towards adoption ($X_{21}$) have been retained after eliminating the trivial variable in the preceding step.

So these variables together have explained 27.9 per cent of total '$R^2$' values of 38.1 per cent. So 2 variables of the total 27 variables merit highest importance in socialization through reinvention ($Y_3$).

**Path Analysis**

The exogenous variables, training received ($X_{27}$) has exerted the highest direct effect on **socialization through reinvention level ($Y_3$)** followed by area under pulse cultivation (Bigha) ($X_5$) and gross return (Rs/ Bigha) ($X_{25}$).

In case of indirect effect on **socialization through reinvention level ($Y_3$)** area under pulse cultivation (Bigha) ($X_5$) has exerted the highest indirect effect on **socialization through reinvention level ($Y_3$)** followed by number of crop diversity ($X_7$) and gross return (Rs/ Bigha) ($X_{25}$).

It is discernible from the table the highest number of variables (22) has routed their substantial indirect effect through the variable, training received ($X_{27}$). So, it could be inferred that these variables have got both substantive and associational properties to characterize the **socialization through reinvention level ($Y_3$).**

The residual effect being 0.6188, it is to conclude that 61.88 per cent of variation in this relation could not be explained.

- **Marketed surplus ($Y_4$)**

**Coefficient of Correlation**

The following independent variables *viz* income (Rs /per capita/annum) ($X_8$), market Orientation ($X_{13}$), productivity or yield (kg/Bigha) ($X_{24}$) have recorded the positive correlation and attitude towards adoption ($X_{21}$). has recorded significant but negative correlation with marketed surplus ($Y_4$).

**Multiple regression analysis**

The $R^2$ value 0.605 being, it is to conclude that even combination of all 27 variables so far, 60.5 % of variability embedded with the consequent variable has been explained

**Step down multiple regression analysis**

The following causal variable *viz* productivity or yield (kg/Bigha) ($X_{24}$) have been retained after eliminating the trivial in the preceding step. So these variable explained 51.7 per cent of total '$r^2$'values of 60.5 per cent. So 1 variable of the total 27 variables merit highest importance in marketed surplus ($Y_4$).

**Path Analysis**

The exogenous variables, productivity or yield (kg/Bigha) ($X_{24}$) has exerted highest direct effect on **marketed surplus ($Y_4$)** followed by farmer's attitude towards Pulse crop cultivation ($X_{14}$) and gross return (Rs/ Bigha) ($X_{25}$).

In case of indirect effect on **marketed surplus ($Y_4$)** gross return (Rs/ Bigha) ($X_{25}$) has exerted the highest indirect effect on **marketed surplus ($Y_4$)** followed by knowledge level of farmer towards cultivation of pulse crop ($X_{15}$) and income (Rs /per capita/annum) ($X_8$).

It is discernible from the table the highest number of variables (22) has routed their substantial indirect effect through the variable, productivity or yield (kg/ Bigha) ($X_{24}$). So, it could be inferred that these variables have got both substantive and associational properties to characterize the **marketed surplus ($Y_4$).**

The residual effect being 0.3948, it is to conclude that 39.48 per cent of variation in this relation could not be explained.

- **Marketable surplus ($Y_5$)**

**Coefficient of Correlation**

The following independent variables *viz* income (rs /per capita/annum) ($X_8$), productivity or yield (kg/bigha) ($X_{24}$) have recorded the positive and significant correlation and family size ($X_3$) has recorded significant but negative correlation with marketable surplus ($Y_5$).

**Multiple regression analysis**

The $R^2$ value 0.774 being, it is to conclude that even a combination of all 27 variables so far, 77.4 % of variability embedded with the consequent variable has been explained.

**Step down multiple regression analysis**

The following causal variable *viz* productivity or yield (kg/Bigha) ($X_{24}$), family size ($X_3$), education ($X_2$) have been retained after eliminating the trivial in the preceding step.

So these variables together have explained 74.3 per cent of total '$r^2$'values of 77.4 per cent. So 3 variables of the total 27 variables merit highest importance in marketable surplus ($Y_5$).

**Path Analysis**

The exogenous variables, productivity or yield (kg/Bigha) ($X_{24}$) has exerted highest direct effect on **marketable surplus ($Y_5$)** followed by Family size ($X_3$) and education ($X_2$).

In case of indirect effect on **marketable surplus ($Y_5$)** area under pulse cultivation (Bigha) ($X_5$) has exerted the highest indirect effect on **marketable surplus ($Y_5$)**

followed by Income (Rs /per capita/annum) ($X_8$) and knowledge level of farmer towards cultivation of pulse crop ($X_{15}$).

It is discernible from the table the highest number of variables (22) has routed their substantial indirect effect through the variable, productivity or yield (kg/ Bigha) ($X_{24}$). So, it could be inferred that these variables have got both substantive and associational properties to characterize the **marketable surplus ($Y_5$).**

The residual effect being 0.2263, it is to conclude that 22.63 per cent of variation in this relation could not be explained.

- **Factor analysis**

**Factor 1**

Factor 1 has accommodated area under pulse cultivation (Bigha) ($X_5$), gross return (Rs/ Bigha) ($X_{25}$), farmer's attitude towards pulse crop cultivation ($X_{14}$), knowledge level of farmer towards cultivation of pulse crop ($X_{15}$), training received ($X_{27}$), family size ($X_3$) which contributed to 10.662 % and has been renamed as **Cognitive motivational.**

**Factor-2**

Factor 2 has accommodated farm size and technology adoption (Bigha) ($X_6$), number of crop diversity ($X_7$), family education ($X_4$), market Orientation ($X_{13}$), education ($X_2$), planning orientation ($X_{11}$), attitude towards discontinuous ($X_{22}$), utilization of cosmopolite sources of information($X_{26}$) which contributed to 7.102 % and has been renamed as **Resource proficiency.**

**Factor-3**

Factor 3 has accommodated productivity or yield (kg/Bigha) ($X_{24}$), age ($X_1$), farmer's attitude towards IPM programme ($X_{20}$), knowledge about fungicide ($X_{17}$), income (Rs /per capita/annum) ($X_8$) which contributed to 6.587 % and has been renamed as **Economic performance.**

**Factor-4**

Factor 4 has accommodated knowledge about weed control ($X_{18}$) (0.585), attitude towards adoption ($X_{21}$), attitude towards rejection ($X_{23}$) which contributed to 6.148 % and has been renamed as **Socialization level.**

**Factor-5**

Factor 5 has accommodated scientific orientation ($X_{10}$), risk Orientation ($X_9$) which contributed to 5.695 % and has been renamed as **Risk management.**

**Factor-6**

Factor 6 has accommodated knowledge about IPM practice ($X_{19}$) which contributed to 5.098 %.

### Factor-7

Factor 7 has accommodated knowledge about insecticides ($X_{16}$), production orientation ($X_{12}$) which contributed to 5.075 % and has been renamed as **Capacity level.**

- **Canonical correlation analysis**

Canonical covariates analysis for this study has been carried out to extract the canonical covariates between two sets of variables, the left side variables and right side variables. it has offered us a dyadic tic interaction between left side variables variable's and right side variables.

It has further gone to form a congenital in group between marketed surplus ($Y_4$) and marketable surplus ($Y_5$) to imply that these two dependent variables have been in clandestine interaction with right side variables viz. the education ($X_2$), farm size and technology adoption (bigha) ($X_6$), risk orientation ($X_9$), scientific orientation ($X_{10}$), knowledge level of farmer towards cultivation of pulse crop ($X_{15}$), knowledge about fungicide ($X_{17}$), knowledge about weed control ($X_{18}$), productivity or yield (kg/ bigha) ($X_{24}$).

On the other hand socialization of pulse enterprise ($Y_1$), socialization through replacement level ($Y_2$) and socialization through reinvention ($Y_3$) have moved together for a clandestine interaction with the following right side variables viz. age ($X_1$), family size ($X_3$), family education ($X_4$), area under pulse cultivation (bigha) ($X_5$), no of crop diversity ($X_7$), income (rs/per capita/annum) ($X_8$), planning orientation ($X_{11}$), production orientation ($X_{12}$), market orientation ($X_{13}$) , farmers attitude towards pulse crop cultivation ($X_{14}$), knowledge about insecticides ($X_{16}$), knowledge about IPM practice ($X_{19}$), farmers attitude towards IPM programme ($X_{20}$), attitude towards adoption ($X_{21}$), attitude towards discontinuous ($X_{22}$),attitude towards rejection ($X_{23}$) , gross return (rs/bigha) ($X_{25}$), utilization/information/cosmopolite ($X_{26}$), training received ($X_{27}$).

### Conclusion

In Indian agriculture, pulse crop is earning an exponential importance because of its tremendous economic and ecological importance. Thus present study evinces that socialization of pulse enterprise has been a complex interplay and analogue between the sets of predictor and predicted characters.

The socialization of pulse enterprise, marketed surplus, marketable surplus, and socialization of reinvention and replacement intervention are being characterize by 27 predictor variables which are mainly socio economic and agro ecological by nature.

The study also responded to the inquiry as to where and how the classical crop production process can be replaced with pulse crop and whereas this replacement will be much rewarding and beneficiary to the common farmer.

The determinants like **gross return, area under pulse crop, training received, productivity of pulse crop and farmer's attitude** are decisively characterizing the socialization process of pulse crop.

If pulse crop remains as an option only for marginal farmer. It can't own grow, it must be an enterprise with both social value and ecological property.

To delineate micro level policy structure, the empirical model of CCA, can help by depicting the fact that the marketable surplus and marketed surplus in pulse crop should be focused conjointly. At the same time effective training intervention on adoption of pulse crop, through replacement and reinvention can offer a boost to this golden crop.

Without pulse in humanity, can survive and it can dovetail both the destiny of soil and society by providing nitrogen to soil and protein to human body.

7

CHAPTER

# LIMITATION OF THE STUDY

There is no research work in this world, which is without any limitation. This is also true for the present study. The most important limitations which have been listed in this chapter are as follows.

I. Bench marking on the status of reigning of '**Technology Socialization Process and market behaviour of pulse enterprise**' has been really a difficult task.

II. There were problems of overlapping in some variables that could have been avoided.

III. Some new scales could have been better if created. Cardinality in measurements has spared.

IV. Since, it is a very nascent idea, more of conceptual inputs could have been better.

V. Schedule complexity because of information overlapping somewhere.

VI. Some of the variables have gone bit homophile in nature. So, this has contributed a fair amount of multi-collinearity characters.

VII. The study does not include gender, class, community and economic category based research.

VIII. Since the concept is complex, it was very difficult to get qualitative response.

IX. Now a day, farmers have become very profession oriented and hesitate to interact and donate information without expectation. That created barriers sometimes to get relevant information unleashed of them.

X. Some more time and more monitory help could have been devoted to generate crystal information.

# BIBLIOGRAPHY


Abbo, S; Saranga, Y; Peleg, Z; Kerem, Z; Lev Yadun, S and Gopher, A, (2009) Reconsidering domestication of legumes versus cereals in the ancient Near East, *Quarterly Review of Biology*.84(1): 29-50

Abdelmagid, S. A and F. K. Hassan, (1996) Factors affecting the adoption of wheat production technology in Sudan". *Quarterly Journal of International Agriculture* 35 (4): 325-337.

Abdul *et al.* (1993) Farmers characteristics affecting adoption of agricultural innovations, *Journal of Rural Development and Administration. xxv* (3): 111 – 113.

Adesina, A.A and Baidu- J Forson (1995) Farmers perceptions and adoption of new agricultural technology: evidence from analysis in Burkina Faso and Guinea. *West Africa Agricultural Economics. 13(1995) 1-9*

Agunga, R. and C. Igodan, (2007) Organic Farmers' Need for and Attitude Towards Extension. *Journal of extension* .45 (6) Feature Articles, 6FEA6

Ahmed, M.M; Preckel, P.V; Baker, T.G and Lopez Pereira, M, (2001) Modeling the impact of technological change on nutrition and marketed surplus, *Agricultural Economics*. 25(1): 103-118

Ajayi, M.T. (2003) Analysis of Mass Media Use for Agricultural Information by Farmers in Nigeria. *Journal of Extension Systems* .19(1):45-53.

Amruta Borate; Zala, Y.C and Darji, V.B, (2011) Analysis of marketable and marketed surplus of red gram in Vadodara district of Gujarat, *Legume Research*. 34(4): 267-272

Arene, C.J (1994) Discriminant analysis of small holder farmer adoption potential and the prediction of extension cost in Nigeria: a comparative enterprise perspective. *Journal of Extension System*. 10 (1): 46–58.

Arunachalam, R, (2011) Spread and acceptance of recommended production technologies in green gram and black gram: a comparative analysis, *Legume Research*. 34(1): 8-13

Asmatoddin, M; Jawale, S.V and Perke, D.S, (2009), Economic analysis of pulses on medium farms in Marathwada region of Maharashtra, *Agriculture Update.* 4(3/4): 262-265

Asrat *et al.* (2009) Farmers' preferences for crop variety traits: lessons for on-farm conservation and technology adoption *Environment for Development Discussion Paper Resources for the Future-RFF-* (09-15): iii + 17 pp.

Bajpai, B.K, (1994) Regional variations in food grains, marketable and marketed surplus in Uttar Pradesh, *Bihar Journal of Agricultural Marketing.* 2(2): 153-163

Bhople, R. S. and S .M.Thakare (1994) Farm information source utilization, adoption and transfer behaviour of farmers; Rural India. *Farm information source utilization, adoption and transfer behaviour of farmers; Rural India.* 57(7): 153-154.

Binkadakatti, J.S; Hanchinal, S.N and Pawar, B.B, (2011) Impact of Krishi Vigyan Kendra trainings in adoption of biofertilizers and bio pesticides practices by pigeon pea growers, *Agriculture Update.* 6(1): 145-150

Biswas, A; Acharya, S.K; Riyang, K and Manashi Saha (2013) Socialization of rubber enterprise: a multifactorial study on livelihood security, *Environment and Ecology.* 31(4A): 1890-1894

Brahmaprakash, G.P; Chandraprakash, J; Ganeshaiah, K.N and Shaanker, R.U, (2004) Pulse yields: feeling the pulse, *Current-Science.* 87(7): 859-861

Burman, R.R; Singh, S.K and Singh, A.K, (2008) Sustainability of pulse based cropping systems in Uttar Pradesh, *Journal of Food Legumes.*; 21(2): 140-142

Burman, R.R; Singh, S.K and Singh, A.K, (2010) Gap in adoption of improved pulse production technologies in Uttar Pradesh, *Indian Research Journal of Extension Education.* 10(1): 99-104

Burman, R.R; Singh, S.K; Lakhan Singh and Singh, A.K, (2008) Extension strategies for increasing pulses production for evergreen revolution, *Indian Research Journal of Extension Education.* 8(1): 5-8

Chand, R, (1991) Agricultural development, price policy and marketed surplus in India study of Green Revolution region, *Agricultural development, price policy and marketed surplus in India study of Green Revolution region.* 128

Charmala, S. and S.M.A. Hossain (1996) Adoption of formal agricultural credit by opinion leaders and other farmers in differentially developed villages of Bangladesh". *Savings and Development.* 20(4): 431-445.

Choudhary, A K and Suri, V.K, (2014) Scaling up of pulse production under frontline demonstration technology transfer program in Himachal Himalayas, India, *Communications in Soil Science and Plant Analysis.* 45(14): 1934-1948

Choudhary, A.K, (2013) Technological and extension yield gaps in pulse crops in Mandi district of Himachal Pradesh, India, *Indian Journal of Soil Conservation.* 41(1): 88-97

Choudhary, J N; Singh, K M and Singh, R.K.P, (1990) Pulses production in Bihar an empirical analysis, *Agricultural Situation in India.* 45(2): 113-119

Deshmukh *et al.* (2007) Knowledge and Adoption of Agricultural Technologies in Marathwada *Indian Res. J. Ext. Edu.* 7 (1): 41-43.

Development Programme Package. *Indian Res. J Ext. Edu.* 9 (2):80-84.

Dolli, S.S and Swamy, B.S, (1997) Impact of adoption of recommended practices on yield of pulse crops, *Karnataka Journal of Agricultural Sciences.* 10(1): 177-182

Fuster, R; Felix, A ; Hegedus Hetze, P de and Gravina Tejera, M.V (2011) Type of subjectivity associated with low technology adoption in fruit culture. *Agrociencia Montevideo.*; 15(2): 158-163

Gautam *et al.* (2008) Effect of Motivation and Knowledge on Decision Making and Adoption of Dairy Practices *Indian Res. J. Ext. Edu.* 8 (2&3) : 55-56.

Haque, M.A; Miah, M.A.M; Ali, A.M and Luna, A.N, (2014) Adoption of mung bean technologies and technical efficiency of mung bean (Vigna radiata) farmers in selected areas of Bangladesh, Bangladesh-Journal-of-Agricultural Research; 39(1): 113-125

Hartwich, F. (2009), Rural innovation networks: from knowledge to development, *Agrarforschung.* 16(6): 186-191.

Hla-Kyi, (1997) Production, consumption and marketing of pulses in Myanmar, *Palawija News.* 14(1): 12-19

Hussain, M.S; Sikder, F.S; Elias, S.M, (1991) Pulse marketing structure and its constraints in Bangladesh, *Economic-Affairs.* 36(3): 145-154

Islam, Q.M.S; Karim, M.R; Ali, M.O and Rahman, M.M, (2002) Adoption of HYV chickpea and the profitability of chickpea cultivation in the High Barind Tract of Bangladesh, *Economic Affairs Calcutta.* 47(2): 100-107

Jadav, N.B. and M.M Solanki, (2009) Technological gap in adoption of improved mango production technology. *Agriculture Update.* 4(1/2): 59-61.

Kad, R.G; Deshmukh, A.N; Mokhale, S.U and Angaitkar, A.G, (2013) Marketing behaviour of pulse growers of Amravati district of Maharashtra, *International Research Journal of Agricultural Economics and Statistics.* 4(1): 115-116

Kantharaju, V; Basavaraja; Teggelli, R.G and Girish, G, (2011) Participatory approaches for productivity enhancement in pigeon pea through improved production technologies, *Environment and Ecology.* 29(1A): 243-246

Kelley, T.G and Rao, P.P, (1994) Chickpea competitiveness in India, *Economic and Political Weekly.* 29(26): A-89-A-100

Khode *et al.* (2009) Adoption of Improved Dairy Cattle Management Practices under Vidarbha

Kunzru, O. N. and Tripathi, H. (1994) A comparative study of adoption of dairy farm technologies between non-members and members of dairy co-operative villages. *Indian Journal of Animal Sciences.* 64(5): 501-507

Lapbim *et al* (2008) Factors that Influence the Adoption of Cocoa Integrated Pest Management by Farmer Field School Graduates in Cameroon *Journal of Extension Systems.*24(1):76-93.

Lee, C. and J.Yang (2000) Knowledge value chain . *J. Manage. Dev.* 19 (9): 783–793.

Moxley, R.L. and K.B.Lang, (2006), The importance of social context influences on new farm technology sustainability: community and sub-community characteristics in Jamaica, *Technology in Society*. 28: 393-406.

Nadarajan, N and Sanjeev Gupta, (2010) Role of classical breeding in improvement of pulse crops, *Electronic Journal of Plant Breeding*. 1(4): 1099-1106

Nath, S.K and Mohapatra, B.K, (2011) Knowledge and adoption level of Rhizobium culture among pulse cultivators of Orissa, *Legume Research*. 34(3): 184-189

Nath, S.K and Mohapatra, B.K, (2011) Knowledge and adoption level of Rhizobium culture among pulse cultivators of Orissa, *Legume Research*.34(3): 184-189

Okkyla, S; Dwijatmiko, S and Kurnianto, E (2014) Path analysis of recording system innovation factors affecting adoption of goat farmers. *Journal of the Indonesian Tropical Animal Agriculture*. 39(1): 58-64

Ommani A.R. and C.Mohammad (2007) Appropriateness of E-learning Based Information Technology to Improve the Productivity of Crops *Journal of Extension Systems*. 23(1): 70-80.

Padaria *et al.* (2009) A Logit Analysis of Bt Cotton Adoption and Assessment of Farmers' Training Need. *Indian Res. J Ext. Edu.* 9 (2):39-45.

Pal *et al.* (2009) Communication Pattern in Drylands of Uttar Pradesh. *Indian Res. J. Ext. Edu.* 9 (1):54-57.

Pant, S.P; Maji, C.C and Sulaiman, R.V (1995) Production prospects and constraints to higher productivity of pulses in Madhya Pradesh, *Policy Paper National Centre for Agricultural Economics*. (2): 48 pp.

Pant,S.P; Maji, C.C and Sulaiman, R.V (1995) Production prospects and constraints to higher productivity of pulses in Madhya Pradesh, *Policy paper National Centre for Agricultural Economics. (2)*: 48 pp.

Patel,V.G, (1999) Economics of pulses production and identification of constraints in raising their production in Gujarat, *Research Study Agro Economics Research Centre,* Sardar Patel University. (110): 58 pp.

Paudel, P. and A.Matsuoka, (2008) Factors influencing adoption of improved maize varieties in Nepal: a case study of Chitwan district. *Australian Journal of Basic and Applied Sciences*. 2(4): 823-834

Pawar, N.D, (1997), Marketed surplus and price spread of green gram: a case study in Parbhani district of Marathwada region of Maharashtra, *Agricultural Marketing*. 40(3): 28-30

Pezeshki,R.G. and M.Masaeli, (2003) Economic factors effective in adopting an integrated campaign in rice stem borer control in Isfahan. *Journal of Science and Technology of Agriculture and Natural Resources*. 6(4): 53-65.

Pothuluru,-C and Yadagiri, P, (1992) Farm output and marketing behaviour of rural poor: some observations, *Agricultural Marketing*. 34(4): 26-29

Prakash, I.J., (1999), Senior women's perception of leisure in India., *Tourism ecreation Research*. 24(1): 82-85.

Prasad, Y E; Rao, M.M and Vijayabhinandana, B, (1993) Analysis of on-farm trials and level of technology on oilseeds and pulse crops in Northern Telangana zone of Andhra Pradesh, *Indian Journal of Agricultural-Economics*. 48(3): 351-356

Prasad, Y.E; Rao, M.M and Vijayabhinandana, B, (1993) Analysis of on farm trials and level of technology on oilseeds and pulse crops in Northern Telangana zone of Andhra Pradesh, *Indian Journal of Agricultural Economics.*; 48(3): 351-356

QiFeng *et al.* (2008) Analysis of farmers' acceptance and response to agricultural technologies in main grain production district - a case from Yushu County in Jilin. *Scientia Agricultura Sinica*. 41(7): 1963-1968

Rahman, M.S; Hossain, M.A; Sarker, M.J.U and Bakr, M.A, (2012) Adoption and profitability of BARI lentil varieties in some selected areas of Bangladesh, *Bangladesh Journal of Agricultural Research*. 37(4): 593-606

Rai, D.P. and Singh,K. (2008) Awareness, Attitude and Training Needs of Farmers about Recommended Practices in Watershed Development Programme *Indian Res. J. Ext. Edu.* 8 (2&3): 89-91.

Ram Singh; Dalip Kumar; Suhag, K.S and Hasija, R.C, (2005) Comparative economics of pulse crops in Rajasthan State, *Environment and Ecology*. 23(Spl-1): 20-24

Ramasubramanian, M; Seeralan, S; Sekar, V and Thomas, M.I, (2010) Cognition and adoption of pulse production technologies by dry land farmers of Tamilnadu, *Legume Research*. 33(2): 102-107

Rao, C.A.R; Rao, M.S; Srinivas, K; Patibanda, A.K and Sudhakar, C, (2011) Adoption, impact and discontinuance of integrated pest management technologies for pigeon pea in South India, *Outlook on Agriculture*. 40(3): 245-250

Rao, P.P and V.G.K Rao, (1996) Adoption of rice production technology by tribal farmers". *Journal of research and ANGRAU 24* (1-2): 21-25.

Rao, P.P; Birthal, P.S; Bhagavatula, S and Bantilan, M.C.S, (2010) Chickpea and pigeon pea economies in Asia: facts, trends and outlook, *Chickpea and pigeonpea economies in Asia: facts, trends and outlook*. vi + 68 pp

Ray, M; Ghosh, P.K and Kajal Sengupta, (2013) Strategies to overcome pulse production constraints in West Bengal, India, *International Journal of Bio resource and Stress Management*. 4(2): 293-297

Rezvanfar *et al.* (2009) Analysis of Factors Affecting Adoption of Sustainable Soil Conservation Practices among Wheat Growers *World Applied Sciences Journal*. 6 (5): 644-651.

Rieu, A. and S. Dahache, (2008) Women starting up in agriculture: from gender socialization to training. *Revue d' Etudes en Agriculture et Environnement*. 88: 71-94.

Rossier, R, (2008), Farm succession: interest and motivation of the coming generation. *Revue Suisse d' Agriculture*. 40(3): 120-122.

Saikia, T.N, (1989) Prospects of changing cropping pattern in favor of oilseeds and pulses production in Assam, *Agricultural Situation in India*. 44(4): 283-284

Sanjay Kumar; Ravinder Singh and Akhilesh Singh, (2014) Assessment of gaps in pulse production in Hamirpur district of Himachal Pradesh, *Indian Research Journal of Extension Education*. 14(2): 20-24

Sarju Narain; Singh, S.K and Lakhan Singh, (2014) Impact of National Food Security Mission-Pulse on chickpea productivity in Hamirpur District of Uttar Pradesh, *Indian Research Journal of Extension Education*.; 14(3): 24-27

Sarju-Narain; Singh, S K; and Lakhan Singh (2014) Impact of National Food Security Mission-Pulse on chickpea productivity in Hamirpur District of Uttar Pradesh., *Indian Research Journal of Extension Education*. 14(3): 24-27

Shakya *et al* (2008) Knowledge Level of Chickpea Growers about Chickpea Production Technology. *Indian Res. J. Ext. Edu*. 8 (2&3): 65-69

Sidhu, R.S; Sidhu, M.S and Singh, J.M, (2011) Marketing efficiency of green peas under different supply chains in Punjab, *Agricultural Economics Research Review*. 24(2): 267-273

Singh, K.M; Singh, R.K.P and Abhay Kumar (2014) Adoption of modern agricultural technologies in Bihar: a farm level study *Environment and Ecology*.; 32(4): 1342-1346

Singh, S.K; Burman,R. R; Chaudhary, R.G; Singh,K. K and Ansari,M. S , (2005) Impact of usable technologies identified under pulse based rain fed agro ecosystem, *Indian Journal of Pulses Research*. 18(1): 60-63

Singh, U.B; Satpal Baloda; Sangwan, N; Gurnam Singh and Sandeep Antil ,(2007) Economics of production and marketing of urd pulse crop in Shivalik foothills of Haryana, *Annals of Agri Bio Research*. 12(1): 63-64

Singh, M (1997) Predictors of attitudinal change of pulse growers: a study with non-progressive farmers, *Journal of Interacademicia*. 1(1): 93-96

Soares, W.L. and M.F.de.S. Porto, (2009) Estimating the social cost of pesticide use: an assessment from acute poisoning in Brazil, *Ecological Economics*. 68(10): 2721-2728.

Subhash Katare and Shrivastava, M.K, (2010) Front line demonstration: an effective technology for increasing productivity of chickpea in Ratlam district, *Agriculture Update*. 5(1/2): 52-53

Surat-Singh; Sarju-Narain (2014) Knowledge and adoption level of IPM practices among tomato growers in Indore District (M.P.). *Indian Research Journal of Extension Education*. 14(3): 125-127

Szmedra, P.I; Wetzstein, M.E and McClendon, R.W (1990) Partial adoption of divisible technologies in agriculture *Journal of Agricultural Economics Research*. 42(3): 20-26

Tchouamo *et al*. (2006) Socio-Economic Impact of Cocoa Crop and Pest Management Knowledge through Farmer Field School (FFS) Approach in Cameroon, *Journal of Extension Systems* .22(2): 01-13.

Thanda-Kyi and von Oppen, (2004) Demand and supply analysis of Myanmar's agricultural export: the case of pulses, *Proceedings of the Annual Research Conference Agricultural Sciences, Yangon, Myanmar,* 10-12- 295-317

Thoke, N.J and Surya Gunjal, (2010) Adoption behaviour of farmers in chickpea production technology, *Agriculture Update.* 5(3/4): 352-355

Uma Sah; Hem Saxena; Narendra Kumar; Singh,S K and Shripad Bhat (2014) Tobit analysis of farmer to farmer diffusion of improved pulse seeds in Bundelkhand region of India. *Indian Journal of Agricultural Sciences.* 84(10): 1254-1261

Upender, M, (1991) Marketed surplus and price spread of groundnut a sample study, *Indian Journal of Agricultural Marketing.* 5(1): 61-68

Usha-Singh, (2001) A comparative study on the productivity level of pulses and other crops, *Journal of Research, Birsa Agricultural University.* 13(2): 211-213

Vergunst, P.J.B. (2009), Whose socialization? Exploring the social interaction between migrants and communities-of-place in rural areas, *Population, Space and Place.* 15(3): 253-266.

Watiez, M., (1994), Psychology and nutrition: study of the process of food socialization., *Medecine et Nutrition.* 30(4): 171-177.

Wijerathna, M; Weerakkody, W.A.P and Kirindigoda, S, (2014) Factors affecting the discontinuation of protected agriculture enterprises in Sri Lanka, *Journal of Agricultural Sciences Sri Lanka.* 9(2): 78-87

Yadav *et al* (2007) IPM in cotton adoption and constraints *Journal of Cotton Research and Development.* 21(2): 230-234.

# INDEX

9 789354 615344

Printed at Repro India Ltd.